Tea Cultivation: Principle and Their Practices

Tea Cultivation: Principle and Their Practices

Shashi Shekhar Prasad

RANDOM PUBLICATIONS
NEW DELHI (INDIA)

Tea Cultivation: Principle and Their Practices

ISBN 978-93-5111-780-3

Published in 2016 in India by

RANDOM PUBLICATIONS

4376-A/4B, Gali Murari Lal, Ansari Road
New Delhi-110 002
Phone : +9111-43580356, 011-23289044, 011-43142548
e-mail: sales@randompublications.com,
info@randompublications.com, randomexports@gmail.com

Reprinted 2026

Type Setting by : Friends Media, Delhi-110089

Preface

Tea is globally one of the most popular and cheapest beverages with major production centres in India, China, Kenya, Sri Lanka, Turkey & Vietnam. The tea industry is one of the oldest organized industries in India with a large network of tea producers, retailers, distributors, auctioneers, exporters and packers. Total tea production in the world has exceeded 4 billion kgs with India producing about 1 billion kg of tea. During 2008 to 2013, black tea production in India increased at a compounded annual growth rate (CAGR) of 1.6% while consumption rose at a CAGR of 2.3%. India's total annual tea production in 2013 is estimated at 1200 million kgs out of which 65%, approximately 850 to 900 million kgs, is produced by the big tea gardens while about 250 million kgs tea is produced by small tea growers with land area ranging from 2 to 20 hectares. Tea export has remained flat over the years due to increasing competition in the global market and declining quality of tea produced in India. Thus the prices in the industry are expected to be stable with domestic consumption expected to be rising steadily.

– Author

Preface

Contents

1

Traditions of Tree Management and Cultivation

INTRODUCTION

Rural men and women in many areas have long been involved in the conservation and cultivation of trees on agricultural lands and forested areas. Until recently, there has been a tendency to discount these indigenous activities. The main focus of forestry efforts has been on management of trees for environmental protection or for industrial timber production. The shift in emphasis towards forestry in partnership with rural people is, therefore, a significant departure from earlier perceptions, policies and practices.

In order to set present community forestry activities, both communally and privately organized, within the context of local spontaneous tree growing, it is necessary to review a number of indigenous tree management and conservation strategies. The relationship between rural people and the trees in their environment is generally complex. Many of the approaches used have been developed over long periods of time. Often they have emerged as responses to increasing, though sometimes subtle, pressures on the local environment. Their underlying function has been to ensure that locally-valued tree species continue to be available to rural people. The extent to which people cultivate and manage trees varies throughout the non-industrial areas of the world.

It depends largely on characteristics of local ecology, patterns of agricultural land-use, cultural traditions, local demands for wood and wood products, tenure rights and economic pressures. In some societies, tree cultivation and management is a major feature of the way of life; in others, it has assumed a peripheral or even a negligible role. Depending on the intensity of the management approach, different strategies will be able to withstand more environmental stress than others. Environmental degradation and the depletion of tree cover is sometimes symptomatic of a lack of comprehensive traditional tree and environmental management systems. In many cases, it has been the result of the breakdown of traditional systems because of intense and interracting pressures.

Introduced innovations may be required where strong indigenous traditions do not exist. In heavily forested areas, and in some regions of the world where alternative resource-use strategies have been pursued, tree conservation and regeneration strategies may be largely absent. Similarly, pressures of poverty, population growth and insecure tenurial rights—among other factors—have sometimes precluded the development or retention of indigenous strategies. Ultimately, rural tree planting programmes should be based on understanding any traditional or existing tree cultivation or management systems which these programmes may supplement, as well as on understanding what has made the introduction of new methods of rural tree management necessary in the first place.

THE IMPORTANCE OF TREES

The world's forests have an undisputed and vital role in sustaining natural and human environments. They protect watersheds, provide habitats for wildlife and help to stabilize otherwise fragile ecosystems. They provide many essential products for rural and urban dwellers. They also play an economic role, with the commercial extraction of timber and pulpwood producing significant amounts of national income and foreign exchange in a number of countries.

Trees Provide Wildlife Habitats

In addition, forests provide a home and a means of livelihood for large numbers of people, both traditional forest dwellers and those who find employment in extracting and harvesting commercial forest products. Forests also play a vital role in agricultural production. Although shifting cultivation has been a cause of the deterioration of much forested land, when it is practiced in an environmentally sustainable way, the natural process of regrowth and regeneration of the forest restores the fertility of fallow land.

Although not as well documented, the role trees play outside established forested areas and reserves is also critical. Trees, dotted about the rural landscape, around houses, along field boundaries and roadsides, and in communal grazing areas, are seldom recorded in the formal statistics of forested lands. But for the majority of the rural population, living away from the immediate vicinity of forested lands, these trees have an even more significant role than the forests themselves.

Therefore, policies and programmes intended to improve access to wood resources and other tree products must be based on a recognition that the trees which will be most useful to these rural people may not be found growing in the forests, but in their own backyards, on smallholdings and on communally held lands.

Useful Trees in the Backyard

In farmlands and grazing areas, trees also play a vital environmental role. They act as windbreaks, protecting crops from wind damage and the soil from erosion. Their shade helps to reduce the temperature of the soil. Tree litter slows down the run-off of rain, thereby protecting the soil and increasing the infiltration of water so that groundwater stores are replenished.

Trees also redistribute nutrients, drawing essential minerals from the subsoil and making them accessible, through their leaf-fall, to other plants. In many countries, tree litter is collected in large quantities for composting and mulching in order to maintain soil fertility. Trees have a valuable social function. They provide shade for people and animals in hot climates, and are sometimes a focal point for family and community gatherings and activities. There are many places where trees are grown and protected for their shade and beauty; sometimes they are treated as sacred.

TREE PRODUCTS

Wood is the most widely used household fuel in non-industrialized areas of the world where supplying energy, in fact, constitutes the greatest demand for wood, far exceeding that of commercial timber. Wood is by far the most important energy source in many countries, accounting for up to 90 per cent of the total fuel used in some of the poorest. In many countries, virtually every rural family uses it for at least some cooking and food processing and for heating. In many cities, charcoal and wood remain the predominant cooking fuels. Some industrial processes such as tobacco and tea curing, brick firing and beer brewing often rely entirely on wood for process heat. Restaurants, tea shops, bakeries and other commercial enterprises add to these demands.

The usual Method of Cooking

However, the dominant role of wood in rural energy supplies should not be allowed to obscure the fact that other traditional fuels are often important. Use of agricultural residues and animal dung is extremely widespread, though since it is poorly understood it is not given the attention it deserves. The result has sometimes been that the case for fuelwood programmes has been overstated because the contribution that other traditional fuels make has not been taken fully into account.

The dominance of woodfuel demand in quantitative terms has also tended to obscure the vital importance of other tree products to rural people. Animal fodder is perhaps one of the most significant of these, especially at certain times of the year when grass and other feed sources are unavailable. In arid areas, trees often provide regular supplies of fodder in the form of edible seed pods and leaves. During times of drought, these become especially important sources of animal feed. A wide range of human foods are also obtained from trees and from forested areas. Some of these are extremely important in preserving the nutritional balance in traditional diets. They include edible leaves and pods, roots, fruits, nuts, honey, insects, and game. Trees may be a source of food condiments, such as spices, and sap from some trees is used to make wine. Mushrooms and other forest funghi are also collected for certain types of dishes.

Trees Can Provide Fodder

Trees produce a large number of what are often referred to as "minor" forest products. The importance of these products should not be underestimated. They make a vital contribution to the needs and general living patterns of large numbers of rural people.

Fig. Mushrooms Can Grow on Trees

Many communities rely on trees to provide fibre to make ropes, mats, baskets, snares, coverings, woven goods and even musical instrument strings. Trees are an important source of many herbal remedies and traditional medicines. Tannins and dyes extracted from tree bark and seed pods are used to cure leather and colour fabrics. Oil from the seeds of certain trees can be used as a substitute for paraffin in hurricane lamps. The leaves and twigs of other trees have good insect repellant qualities, are used to stun fish or serve as natural livestock dewormers. Some resins can be used as glues.

Basket Making—A Family Business

Agricultural implements, bullock carts and boats are often crafted out of wood, and certain varieties of wood are highly-valued for their tool-making qualities. The trunks of some trees are hollowed out for water storage. African woods are used to make camel bells.

A Boat Crafted from Wood

Trees also supply a variety of commercial construction materials. Building poles, for example, are widely needed for a variety of purposes. In rural areas, they are used in construction of many traditional types of dwellings; in cities, poles are used by the poor in constructing low-cost housing, and larger poles are extensively used for scaffolding and for props by the building industry.

Harvesting and distribution of these tree products is an important income-generating activity in many areas. Charcoal production and fuelwood vending, for instance, are a vital source of income for many poorer rural households. Other jobs are generated by pitsawing, operation of sawmills, woodworking, tree farming, and the gathering and selling of fruits, timber, resins, gums and other forest products both for cottage industries and for larger scale commercial concerns. The significance of small-scale rural enterprises which process forest materials is gradually becoming better documented.

In a recent survey, FAO has shown that such enterprises are often one of the largest sources of off-farm employment and income. Further work is now in progress to determine the key characteristics of such enterprises and to formulate ways to enhance their economic contribution.

The role of trees in the patterns of rural living is thus both complex and diverse. Focusing on one aspect, even such a major one as fuelwood, at the expense of others can be gravely misleading. So much of the fabric of rural domestic and farming life relies on trees that any diagnosis of the problems arising from the depletion of tree cover which does not take into account the full complexity of this dependence is likely to be inadequate.

THE ROLE OF TREE MANAGEMENT AND CULTIVATION

Almost everywhere, a certain standing stock of different types of trees, whether deliberately cultivated or allowed to grow naturally, has been recognized as necessary by farming communities. Even in nomadic pastoralist societies, trees have traditionally played a variety of essential roles. Though pastoralists may rarely have planted trees, their traditional way of life was such that they did not usually deplete the supply in the territories over which they ranged with their flocks. Indeed, the grazing animals helped maintain the tree base by dispersing tree seeds over wide areas.The many products and benefits which rural people derive from trees reflect detailed and sophisticated knowledge about their immediate environment. The assumption that traditional communities are unaware of the benefits provided by trees, and therefore need to be educated about the immediate consequences of the depletion of tree cover, is rarely accurate.

Local impacts such as the loss of fodder, shade, fruit, and other benefits are obvious. Although rural populations may not have a clear understanding or perception of the long term consequences of deforestation—particularly downstream consequences—their ability to name and distinguish a large number of species and to describe their characteristics demonstrates awareness of trees and the role they play in their own lives.

In some cases, rural silvicultural systems are highly sophisticated with considerable numbers of trees planted and well-developed techniques used for managing and harvesting them. Elsewhere, managing tree resources is more passive and relies on conservation and natural regeneration. Stability of the system rests on the fact that population pressures are low and that the forest's capacity for regeneration is great enough to offset any damage done by the utilisation practices of rural people. Where traditional societies have remained stable, they have generally been able to maintain the productive role of the tree resources on which they depend. Though traditional tree management strategies might slow or even stop the processes of environmental deterioration, the primary focus has usually been on the utility value of trees for household or community use. Some practices have resulted in developing elaborate agroforestry systems, such as home gardens, which have incorporated indigenous trees into sustainable production systems. Others have been more modest in scope and effect, based on the desire to retain at least some valued trees conveniently near the household.

TRADITIONS OF TREE PROTECTION AND MANAGEMENT

Restricting access to trees is one means by which individuals, households or kinship groups may assert exclusive rights to them. The use of the baobab in southern Niger, for instance, is defined by very old traditions which specify strict proprietary rights. In Sudan, palm trees may be subject to a complex system of fractional ownership defined by traditional laws of inheritance. In Western Sumatra, the decision to cut a valuable tree is made by the extended family.

A Baobab Tree—Protected by Tradition

Some communities have long protected specific trees because they provided a focal point for the community, occasionally because of their religious significance. In Nepal, formal management systems were developed by some communities over a period of centuries. These systems defined specific users' rights to valued products from trees growing on common lands. The systems were as much a response to distributional demands as they were to growing scarcities. As some of the older systems broke down, visible recent increases in the rate of forest destruction prompted some communities to establish new systems.

Other groups of people with common interests in tree resources have also been responsive to the threat of growing scarcities. In the highlands of Guatemala, professional woodworkers have been a particularly strong force in pushing for tree conservation efforts.The Chipko Movement in the Himalayan region of India is a communal effort with active leadership from women, which has relied on the Gandhian technique of non-violence to protect trees from destruction by commercial timber concerns. The Karen people of Thailand have customarily tried to contain swidden plot fires. Among particular tribal groups in Kenya, honey collectors are obliged to prevent fires when smoking out bees; and in some areas of India, cutting down a tree may be considered unethical, especially if it yields products which are of use to the community. Among the Bora Indians of the Peruvian Amazon, there is a recognition that their system of shifting cultivation must be managed in a way which both reduces soil erosion and favours particular trees in secondary vegetation

In addition to active tree conservation efforts, some local land management strategies have consciously matched demands on the land with its carrying capacity. In these areas, it is clearly recognized that overgrazing results in environmental degradation; the size and grazing patterns of livestock herds are therefore kept within environmentally acceptable limits. Other strategies for managing tree resources have involved protecting and cultivating naturally germinating seedlings. Cultivators may leave certain desirable tree seedlings when weeding and even build barriers around them as a protection against grazing livestock.

In parts of southern Mexico, farmers tolerate and protect indigenous leguminous trees, such asProsopis, which provide edible pods, shade and enhance soil fertility. Cultivators in southern Nigeria recognize the superiority of certain species in restoring soil fertility to fallow plots and encourage them to dominate the bush.

A Prosopis Providing Fodder

A recently emerging management strategy has been to restrict peoples' access to trees previously available to them. This approach is usually associated with changes in systems of land tenure and may be a response to worsening wood scarcities. In central Kenya, the collection of wood or of other products from trees growing on private lands has increasingly required the permission of the property owner, although until recently, trees and their products have been free goods.

COPPICING AND POLLARDING

Coppicing and pollarding are techniques which can be used in managing certain types of trees. Coppicing involves cutting the tree down to its stump and allowing it to regrow; it normally sends up a number of shoots instead of the original single stem. Pollarding involves cutting off the crown of the tree, leaving it to send out new branches from the top of the remaining stem; this also has the advantage that the new shoots are high and therefore more apt to be protected from animal and fire damage. Vertical pollarding is the close pruning of branches along the stem. Regrowth after coppicing and pollarding is vigorous because the tree's root system has already been well established.

A variety of different pollarding, coppicing and pruning techniques have been reported in many parts of the world, for instance in Bangladesh, Burkina Faso, the Philippines and Rajasthan, India (Douglas, 1981; Wiersum and Veer, 1983; Ben Salem and Tran van Nao, 1981).

Pollarding—A Rational Tree Management Practice

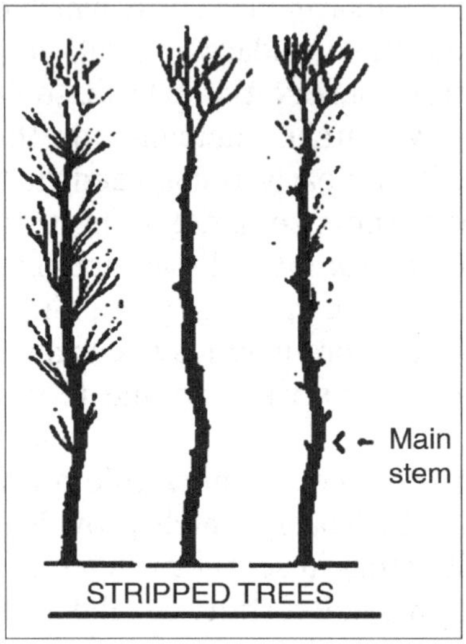

Fig. Vertical Pollarding

It has been reported that in the highlands of Kenya the pollarding of Grevillia growing on agricultural lands is common. Intense pollarding of these trees may be carried out 15 or 20 times over a period of 50 years. The trunk will continue to widen, and the stem will increase in height unless this is deliberately prevented by pruning at the top. Whenever the farmer decides that it is large enough or that he needs the money, the trunk is felled and is sold for timber (Poulsen, 1983). What these techniques all have in common is that they permit a sustained yield of wood or fodder over a long period of time. The total lifetime contribution of a tree which is used in this way can be considerably greater than the volume it will produce if it is simply allowed to grow and is then felled. The fact that coppicing and pollarding techniques are widely employed by farmers has often been overlooked. In some areas, however, it is clear that their use enables an overwhelming proportion of the fuelwood or timber needed by households to be obtained on a sustainable basis from trees growing on agricultural land. This has important implications for the design of programmes to maintain increased fuelwood supplies.

TRADITIONS OF TREE PLANTING AND CULTIVATION

In many countries, rural people traditionally plant trees for a multiplicity of household uses. In Bangladesh, one of the most densely populated of all countries, it was found in 1983 that on average each household had planted or naturally regenerated 68 trees, of which 16 had been established in the previous year. In the area of Fatick, Senegal, it has been found that virtually all households had planted trees.

In Panama, fruit trees are planted on almost every small farm (Jones, 1982b). Nearly half of the farmers interviewed in the Valle Occidental region of Costa Rica said that they had planted trees as windbreaks. In Peru, spontaneous private tree plantings account for around 30 per cent of all trees planted, even though the government only sponsors large-scale reforestation. A survey in the hill area of Nepal showed that on the average each household owned 28 trees, around a third of which had been planted and cultivated (Campbell and Bhattarai, 1983). Of rural households in the Kakamega District of Kenya, nearly 40 per cent maintain small nurseries and nearly 80 per cent have planted trees on their land.

Perhaps surprisingly, there are few reported cases where rural people have spontaneously planted trees specifically for fuelwood, except where they intend to sell it. In Kenya, for example, it has been found that people plant trees for fruit, to provide shade or ornament, to create windbreaks, or to mark out boundaries In Malawi, large numbers of farmers plant trees, mainly for poles to be used by the family. Only 15 per cent of the people interviewed in a study of tree planting practices planted trees for fuelwood (Energy Studies Unit, 1983). In spite of heavy demands for rural energy supplies, Indians living on Bolivia's Titicaca plateau felt that trees were too valuable to be used as fuel; they used them mainly for house posts and for other objects of material culture.

Although trees are seldom planted specifically for fuelwood, there is still an awareness that they will help to supplement available energy supplies for the household. Trees which are planted for use as timber or as construction poles, for instance, must be trimmed in order to make sure that they grow straight; fruit trees must be pruned occasionally if they are to produce better quality fruit. These cuttings can be used as fuel. Building poles which have been replaced at the end of their useful life are also used as fuel.In a few areas, however, there are established traditions that involve the planting of fuelwood for commercial markets. Around the Indian city of Madras, casuarina plantations were originally established in the late 19th century to provide wood for the railways, but were used for household energy when the railways were converted to coal.

During the Second World War, when acute fuelwood shortages developed, local farmers seized the opportunity to plant trees for the urban fuelwood market. The practice still continues, and similar plantations also exist around other South Indian cities. In several areas of Java, farmers have responded to increased commercial fuelwood demands by planting large areas of Calliandra (National Research Council.

Trees have also been cultivated to provide numerous products for other specific markets. Tree products such as gum arabic, rubber, coconut, dates, palm oil, coffee and tea are vitally important elements in the economies of many developing countries. Cultivation of these trees is not confined to large plantations, as they also provide cash incomes and a means of livelihood for large numbers of smallholders and limited resource farmers.

Rubber—Trees for a Specific Market

Commercial timber species are also cultivated by farmers. For many years the match-making industry in southern India has been based largely on smallholder tree growing. Some companies distribute free seedlings to ensure a steady future supply of wood. Sometimes trees are seen as a method of long-term insurance. In parts of Latin America, it is common for farmers to plant a few trees around their dwellings to be cut and sold for timber when money is needed. In Turkey, trees are traditionally planted to celebrate the birth of a female child as a kind of "down payment" on her wedding.

TRADITIONAL AGROFORESTRY SYSTEMS

In many areas, rural people have combined tree growing with a variety of agricultural and pastoral activities on the same plot of land. In general, the most widespread benefit from this type of systematic tree-pasturage-crop combination is the soil enriching effect of the trees. An additional benefit is obtained from the protection against erosion. The total productivity of the land is also increased by the fact that these systems permit a supplementary or complementary use of different layers of the soil and of the space exposed to sunlight above the surface.

The types of practices and their productive outputs vary widely. Cultivators in arid zones of Rajasthan, India, intercrop fodder and grain crops with Prosopis cineraria. In the case of the failure of their other crops, Prosopis becomes a main source of fodder; leaves and seed pods are stored for livestock in anticipation of lean periods. The wood is used for charcoal and fuelwood and for making agricultural implements. Cordia alliodora is used as a canopy tree over coffee and cacao in the humid lowlands of tropical America. Its predominance is so great that it has been estimated to be the third most widely grown tree in some areas, even though it is not recorded in any statistics of tree plantations (Budowski, 1983). In much of the Sahel, naturally germinating Acacia albida is allowed to grow in fields in order to improve the soils (Weber and Hoskins, 1983).

Acacia Albida—A useful Tree

Where population densities are low and land is abundant, fallow periods can be long enough to allow shifting cultivators to practise a successful type of agroforestry. By leaving some trees in swidden plots and encouraging the growth of soil-enriching trees and plants, some shifting cultivators have played a direct role in maintaining local tree cover and speeding up the fallowing process. Among Lua cultivators in Thailand, 84 varieties of plants and trees were found in swidden plots, including 70 that provided food and 13 species with medicinal uses. Swiddeners in Sumatra leave fruit trees and hiving trees for bees (Pelzer, 1948). Shifting cultivators in the Peruvian Amazon have been found to favour certain commercially valuable timber species in fallow plots (particularly cedar) in antiçipation of substantial cash returns for their children. Among the most elaborate systems of indigenous agroforestry are the "home gardens" of Southeast Asia, Latin America and Africa.

Traditional Agroforestry—the Home Garden

These gardens usually have a multi-layered mixture of a large number of food, fodder and wood producing species grown in close association. They are commonly grown in small plots located close to individual dwellings and are very carefully tended; frequently, they are also used for keeping poultry and small animals. Basically, home gardens imitate or recreate the multistoreyed structure and species diversity of forests; this allows a simultaneous combination of perennial and annual crops in a small area. The diversity of species which are grown in home gardens results in a wide range of products. Because the crops have different biological cycles, a household is usually able to harvest some produce on a daily basis, even if in small amounts. On plots as small as a tenth of a hectare in Central America, perhaps 25 or more varieties of food plants and trees have been found, including coconuts, papaya, bananas and coffee (Wilken, 1978).

A study of home gardens in Indonesia, which cover around 20 per cent of the arable land on the island of Java, revealed 37 species of fruit trees, 11 species of food-producing plants, 12 medicinal species, 21 herb species, 18 vegetable species, 45 species of decorative plants and 47 species of plants used for fuelwood and for construction—all growing at a single site A number of swamp or moist-land cultivation systems incorporating trees have much in common with home gardens. Thechinampas system has been used in parts of Mexico for hundreds of years. Raised platforms are constructed, and sediments from the bottom of swamps or from specially constructed reservoirs are used to grow a wide variety of annual and perennial crops. Fruit and other trees which provide shade, support for vines and other products are planted along the edges of the platforms or are interspersed with plants growing on the platforms themselves. A similar practice is found in the deltaic plains of Bangladesh where villages rest on groups of mounds as protection against seasonal flooding. Multistoreyed plants, shrubs, bamboos, palms and other trees are grown in these fertile alluvial soils.

On the lower Tana River in Kenya, farmers plant a number of plots with different annual and perennial crops in order to minimize their environmental risk. Although a few plots might fail during a given planting season, because of the different agro-ecological conditions in each plot and the variety of crop demands, the likelihood of total crop failure is minimized. Usually, plots are set aside specifically for fruit, fuelwood and local construction materials. Agroforestry systems have often developed in areas of high population density as a response to land shortage. If, however, landholdings become too small as a result of demographic, economic or political pressures, the limits of adaptability of the system may be reached and short-term solutions adopted. Trees which had been grown in a complementary relationship with other crops may be uprooted in order to make room for the crops necessary for subsistence. It has been found that in areas where average landholding sizes have substantially decreased, farmers often revert to the production of just a few staple crops such as cassava (Wiersum, 1984).

TREE MANAGEMENT PRACTICES IN PERSPECTIVE

Traditional strategies for managing tree resources are dynamic by nature. They have developed as responses to particular situations, reflecting a variety of cultural, social, economic, political, ecological and demographic factors. Where they have survived successfully, they have often been able to accomodate the introduction of new agricultural crops, the growth of populations, the expansion and contraction of market opportunities for particular crops and other factors. However, the fact that rural people have, in the past, been able to manage their tree resources effectively does not necessarily mean that they can continue to do so. Increasing economic, demographic and social pressures have contributed to the breakdown of traditional tree management practices in many areas. The more passive systems which rely principally on the natural regenerative capacity of forests and woodlands have been particularly vulnerable; in some areas they have been completely overwhelmed. In order to understand the context for the successful introduction of rural forestry innovations, it is necessary to understand why people are constrained from planting and managing adequate numbers of trees in the first place. Equally and perhaps more important is an understanding of why natural forests and trees growing on and around farmlands, which formerly were able to meet the diverse tree-based needs of rural people, are no longer able to do so. Table Some examples of prominent agroforestry systems and practices in the developing countries

PLANT BIODIVERSITY

SOCIO—ECONOMIC, ETHICAL AND LEGAL IMPLICATIONS

Some 40 per cent of the world's market economy is based on biological products and processes. In the rural communities of Africa, Asia and Latin

America where the majority of the world's people live, the dependence on biomaterials can run to over 90 per cent of human survival requirements.

In an increasingly urbanized world, it is difficult for those of us inside city gates to remember that more than half of the food humanity consumes is bred and produced by the people who eat it, and that eight out of ten members of the human family turn to community healers and medicinal plants for protection from illness.

WORLD HEALTH ORGANIZATION PROGRAM

About ten years ago, the World Health Organization launched its programme of 'Health for all by the year 2000.' Right from the outset of the programme, WHO realized that any programme adopted in any of the minimally industrialized countries would have no impact if it did not take into consideration the development and integration of traditional medicine into the primary health care programmes of these countries. WHO has estimated conservatively that between 60 and 90 per cent of the population of the non-industrialized countries rely on medicinal plants to meet their health care needs, either totally or partially. With the scarcity of doctors and paucity of hospitals and clinics, the large majorities of these populations have to rely on sources other than allopathic medicine for their health care. For example, in Ghana there is one traditional doctor for approximately every 400 people, while the ratio of allopathic doctors to patients is 1 for 12,000.3

Various reports from the United Nations (UNCTAD and GATT) have indicated that 33 per cent of drug products in the highly industrialized countries are derived directly from higher plants; most of these are tropical plants growing in equatorial countries. Lower plants and microbes produce another 27 per cent of drugs found on the market.4 Indeed, more than 67 per cent of the world's plant species—at least 35,000 of which have potential medicinal value—originate in non-industrialized countries of the 'South'.5 The market for plant-based drugs is growing every year throughout the world. These drugs now account for over $US 50 billion of the total worldwide drug market now totaling over $US 173 billion.

The estimated manufacturer's price value of medicinal materials derived from plants obtained from the developing countries of the South is expected to reach between 35 and 47 billion US dollars by the year 2000.6 Of the total world pharmaceutical sales of $US 172.56 billion, as of 1989, Africa accounted for only 2.1 billion US dollars. In fact, the highly industrialized countries—which constitute only 20 per cent of the total world population of now over 5 billion—consume 80 per cent of all pharmaceuticals produced in the world. So, how do Africa's 500 million people, for instance, take care of their health if their total share in world pharmaceutical consumption is only 1.2 per cent (US$ 2.142 billion)? Indeed, the total for all developing countries is only 15 per cent of worldwide pharmaceutical purchasing.

Whatever means or method used by this section of the world's population ought to be fully recognized in any formulation of health policies or programmes aimed at 'health for all' by the year 2000.7 The international community has a clear duty to ensure the full conservation of whatever alternative sources of health care are currently available to the remaining 80 per cent of the world's population normally classified as belonging to the 'South'.

Right from the beginning of the WHO's programme, the organization recognized the peculiar circumstances that obtain in less industrialized countries with respect to traditional medicine and health care delivery. Both WHO and UNICEF recognized that in view of the widespread use and acceptability of traditional medical practice, no impact will be made on Africa's overall health care status without due recognition, development and integration of traditional medicine into the primary health care delivery system of each country or region. This recognition led to the WHO/UNICEF 1978 conference in Alma Ata, USSR, at which the participants resolved and specifically urged member states to:

- Initiate comprehensive programmes for the identification, evaluation, cultivation and conservation of medicinal plants used in traditional medicine;
- Ensure quality control of drugs developed from traditional plant remedies by using modern techniques and applying suitable standards and good manufacturing practices.

This conference was followed in 1988 by another in Thailand, at which the Chiang Mai declaration to 'save plants that save lives' was made. This declaration formally brought the rational and sustainable use and conservation of medicinal plants into the arena of public health policy.

The Alma Ata declaration of 1978 was followed subsequently by various policy decisions of the Inter-African Committee on Medicinal Plants and African Traditional Medicines of the Organization of African Unity (OAU). These policy decisions were aimed at implementing WHO's recommendations. Several African countries including Ghana have initiated programmes aimed at maximizing the contribution of herbal medicine to their primary health care delivery nationwide.

These programmes have raised some very pertinent questions and problems of considerable importance for Africa particularly, and for the 'Third World' in general. These include questions that are socioeconomic, ethical, legal and environmental.

Development of medicinal plants relies very heavily on the knowledge carried by indigenous peoples and rural societies. In recent times this has raised concerns about equitable sharing of the benefits of such knowledge and the intellectual property ('IP') rights of these indigenous rural communities. Securing such rights and protections entails specific dilemmas and policy implications, which I will examine throughout the rest of this essay.

Socioeconomic Considerations

World trade in medicinal plants accounts for about 30 per cent of the total drug market. This percentage excludes plants used as raw material sources for the essential oils required to manufacture cosmetics, food additives and for other non-medicinal purposes. Most of these medicinal plants come from less industrialized countries that supply them as cheap raw materials to the multinwe2 ational pharmaceutical manufacturers in the more industrialized countries. Of 76 compounds obtained from higher plants that are present in US prescriptions, only 7 per cent are commercially produced by total synthesis.8 In 1976, imports of medicinal plants worldwide were estimated to be $US 355 million.

This rose to $US 551 million by 1980—an increase of over 60 per cent within four years.9 But when these medicinal plants are processed into suitable dosage forms as safe and efficacious drugs, they are sold to the public at prices far beyond the affordable range for the majority of people in the relatively poor countries. Yet it is the botanic resources of these same countries that make possible the manufacture of these drugs for commercial markets. A typical example is the Madagascar rosy periwinkle Catharanthus Roseus, a plant indigenous to Africa that has been used for several years in East Africa for the treatment of cancer and diabetes.

Research on this plant yielded two very important anti-cancer drugs: vinblastine and vincristine. Although the plant is native to Madagascar, it was the US pharmaceutical firm Eli Lilly that commercialized the drugs from germplasm obtained from the Philippines and Jamaica.10 Vincristine is now the drug of choice for the treatment of childhood leukemia, and it has an annual retail value of over $US 150 million. Vinblastine is used for treating Hodgkin's disease, a serious ailment of unknown origin resulting in enlarged lymph glands, increased numbers of lymphoid tissue cells (hyperplasia) in the spleen, liver and other organs, as well as anemia. Before the discovery of vincristine, the survival rate of children with leukemia was about 20 per cent. Now the success rate for this drug is over 80 per cent. But a 5mg ampoule of vincristine sulfate injection costs about $US 80, making the drug virtually inaccessible to most leukemia children outside the industrialized world.11

Maytansine belongs to a class of compounds called the ANSA Macrolides, first discovered from microbes and now found to occur in higher plants. It was first isolated by Kupchan in the US from the plant Maytenus serrata obtained from Ethiopia. Maytansine showed promising clinical results as a potential antileukemic drug, but died of cancer before his work could be published. Later, higher yields of the compound were found in a Kenyan species, Maytenus buchananii, and a South African plant Putterlickia verrucosa. The compound is active against various strains of lymphocytic leukemia and melanoma, and has been on clinical trials for some time. It is interesting to note that one of the

Maytenus species has been in popular use for almost 50 years in a South African herbal mixture for carcinoma and sarcoma.

But alas, as in the case with vincristine, if maytansine finally gets on the market neither Ethiopia, Kenya nor South Africa will get the millions of dollars accruing from its sales. (Nor is it likely that an average citizen needing the drug in any of these source countries will be able to afford it). Plants are not patentable; but synthetic routes or isolation methods are.12 Similar examples abound.

Unfortunately, even though our herbalists can continue to provide some relief for cancer patients by administering the crude plant extracts of Maytenus andCatharanthus, their knowledge, which originally led to the discovery of the landmark drugs just described, cannot be rewarded in the same way as multinational drug companies get their rewards and profits. This means that the non-industrialized countries are not deriving any economic benefit from the exploitation of their nations' medicinal plant resources. Moreover, as noted already, the approach adopted by the Western world for production and distribution of drugs from plants is far from satisfactory for the majorities living where these essential resources exist. Therefore, in formulating policies for the exploitation of our medicinal plants, policy makers will have to consider how best to manage our natural resources for the full benefit of our citizenry.

Ethical Questions: Intellectual Property Rights of Indigenous Peoples

The ethical issues involved in medicinal plant research and utilization can be divided into two main categories, namely:

1. The relationship between the researchers based in highly industrialized countries and the host countries whence the raw materials are obtained;
2. The relationship between research and development personnel and commercial concerns.

With every passing day large quantities of raw materials are exported to industrialized countries, either for research aimed at producing new drugs or for direct production of drugs for sale. At the time of the 1994 Earth Summit in Rio de Janeiro, the US National Institute of Health and USAID launched a 'drug discovery' project in the less-industrialized countries of the South.

The strategy involved making use of the "wealth of knowledge held by traditional countries".13 Through its departments of Chemistry and Botany, the University of Ghana has been part of this project over the past five years, in collaboration with the National Cancer Institute (NCI) and the Missouri Botanical Gardens. I shall describe our experiences with this project in due course. Similarly, the Shaman Pharmaceuticals Company has pioneered new approaches to working with rural communities that seem to be bearing commercial fruit. For example, a cooperative agreement has been made with the Consejo Aguaruna y Huambisa in Peru. About half of the 400 species collected through this agreement have shown some medicinal potential, and two drugs are now in clinical trials. By working with community innovators, the researchers' efficiency in screening plants for medicinal properties has improved considerably —by more than 400 per cent. Notwithstanding what transpired at the Rio Earth Summit, it appears that now the crucial contribution of indigenous and rural communities to innovations in the conservation of the world's biodiversity has finally been recognized, but is yet to be fully understood.

As observed by the 'Crucible Group', the fact that "indigenous peoples inhabit the most diverse fields and forests of the world is sometimes viewed as both coincidental and unfortunate. That a correlation could exist between the uses made by people of biological diversity and the availability of that diversity is seldom considered".

Many new drugs discovered from plants have originated from information obtained from a local informant about the traditional or ethnobotanical use of the plant. But once the drug is commercialized, the source of this original piece of information gets neither benefit nor financial compensation for sharing his knowledge and expertise. There is no protection of the individual's intellectual property, however rudimentary such protection might be. This raises not only ethical, but also legal questions which I shall deal with later. Moreover, if the particular plant is successfully commercialized, the country that is the main source of the plant will find itself exploited as a supplier of cheap raw materials. A typical example is the case of pilocarpine, an alkaloid from the Brazilian Pilocarpus species, used for treating glaucoma.

In 1989 the USA sold $28 million worth of pilocarpine. Yet the Brazilian producer was paid only 28 cents per kilogram of Pilocarpus. In 1962, 808 tons of this plant were exported from Brazil at a price of 22 cents ($US 0.22) per kilogram, while 1,290 tons were exported in 1970 at 36 cents per kilogram. The question arises: is it ethically or morally right for research agencies from 'developed' countries to go to 'developing' countries, obtain valuable information crucial for research and manufacture of a precious drug that is likely to be inaccessible to the populations of the 'developing' countries, without the latter receiving any meaningful economic benefits for originally providing the knowledge that led to the discovery or development of that drug?

As a corollary: will it be ethically or morally justified for 'developing' countries to protect their traditional herbal heritages by placing a total embargo on the exploitation of these medicinal plants? The Organization for African Unity (OAU) has regulations that forbid export of medicinal plants in commercial quantities without the explicit permission of the host government. Research cooperation is likewise supposed to be regulated and monitored. But except for a few countries, these regulations are totally ignored. For example, several tons of Voacanga africana, a plant that serves as a raw material source for anticancer drugs, are exported yearly to the USA for pharmaceutical industries there. There is a plant called Thaumatococcus danielli that contains a sweetener believed to be about 4,000 times sweeter than sugar. In fact it is supposed to be the world's sweetest natural substance: one kilogram is as sweet as four tons of sugar. Moreover, being a protein, it does not pose the same high calorie problems that sucrose poses. Tate and Lyle of Britain have shipped literally tons and tons of this plant out of Ghana and Nigeria, and developed it into the sweetener 'Talim' that is suitable for diabetics. What benefit is Ghana or Nigeria gaining from all this enterprise? Correlatively, would it have been prudent or ethical for Ghana to place an embargo on the exportation of this plant, thus potentially impeding its development for the general good of mankind?

A US-based company [name withheld], in collaboration with a Swedish firm [name withheld], is currently working on methods for producing another protein-based sweetener from another plant, Synsepalum dulficicum (the sweet berry), again using material obtained from Ghana. This plant also occurs in many other African countries. So there arises the ethical question of how to determine which country should be the beneficiary as the original source of the raw material, in the event that any compensation is eventually awarded by these two companies.

Patent Laws and Intellectual Property Rights

In every country, there are certain legal requirements that govern all prescription drugs—their manufacture, wholesale and retail distribution, dispensing procedures and usage. However, when it comes to phytopharmaceuticals, certain peculiar legal problems arise. This is especially so in most African countries where there is no official legal framework for the incorporation of traditional medicines into the health care delivery programme.

In the European Community (EC) countries generally, all phytopharmaceuticals are subject to general drug regulations. Like synthetic drugs, all phytopharmaceuticals must comply with criteria of quality, safety and efficacy. Their pharmacological, toxicological and clinical profiles have to be scientifically established through standardized tests and deposited with the appropriate licensing committee before permission can be granted for the drug's distribution and sale. For example, according to French law only the pharmacist is licensed to retail herbal medicines, with few exceptions. There are specific legal definitions of what constitutes a vegetable drug.17 In the EC, a medicine is defined as any substance or combination of substances presented for treating or preventing disease in human beings or animals, with a view to making a medical diagnosis or to restoring, correcting or modifying physiological functions in human beings or in animals.

An herbal drug will therefore be considered as a medicinal product if it is:

- Presented for treating or preventing disease in human beings or animals;
- Administered with a view to restoring, correcting or modifying physiological functions in human beings or animals. Hence, for use as drugs, herbal preparations are legally required to be tested for contamination by foreign materials, pesticides, foreign plants, parasites and bacteria. To get a license for marketing in the EC, a drug must meet the basic standards of quality, safety and efficacy.18 But in Africa very few countries have put in place any legal framework to regulate the production of herbal drugs and the practice of herbal medicine.

India does have a comprehensive law. Ghana is only just about to submit a draft law to Parliament for approval. Meanwhile, Ghana has set up a unit within the Ministry of Health that acts as a quasi-regulatory authority for the practice of herbal medicine. But this unit is yet to be given appropriate legal backing to enforce its regulatory and safety measures.

One of the difficulties bedeviling any attempt at integrating traditional medicine into primary health care has been the secrecy with which each herbalist guards his own preparations for any particular ailment. I have already mentioned that many new drugs developed from a plant source originate initially from the folkloric knowledge of the local herbalist. This was the case for aspirin, quinine, vincristine, physostigmine, pilocarpine, morphine, codeine, atropine and other essential drugs. Now, the problem is that there are no patent laws regarding such knowledge. Normally one cannot patent a plant.

One can only patent a particular formulation, or a substance isolated from a given plant. Suppose an herbalist tells a scientist about a plant used for treatment of, say, hypertension. If the scientist is then able to isolate the active ingredient or to formulate even a crude herbal preparation that is thoroughly tested and accepted, then he, the scientist, can patent the drug or the preparation. But the herbalist cannot patent the plant material, because this is considered to be 'God's universal property.' Moreover, that same plant may exist in so many different countries and may have been used for years for the same purpose by many people independently.

So the herbalist has no other alternative but to keep his knowledge close to his bosom. This has serious implications. Suppose a policy is formulated compelling the herbalist to subject every preparation to scientific quality control, to test it for efficacy, toxicity, and so on—as is the norm in Europe and America, and increasingly now in African countries. Then the herbalist will be compelled by law to reveal the name of the plant, the source, his mode of preparation and so on, to the licensing authority. The protocol of secrecy exists even in highly industrialized countries, but only during the time that a drug is being developed by a pharmaceutical firm. Once patenting is completed and the legal protection is secured, all the information about a drug can become public knowledge and can even be published in professional journals without threatening the credit and benefits due its discoverer.

So the question remains of how to resolve this thorny problem of effectively protecting the intellectual property of our local herbalists, whose knowledge and expertise have been handed over from one generation to the next in accordance with established traditional norms.

BIODIVERSITY AND INTELLECTUAL PROPERTY

In recent times, various organizations have tried to draw the attention of the international community to the need for formal recognition and protection of the intellectual property rights of indigenous peoples, particularly the rights of farmers, herbalists and custodians of a nation's plant genetic resources. In highlighting this need, these organizations have not lost sight of the interdependence of all regions of the world in accessing some of the most important foods and medicines of the biosphere. Governments and industries have recognized this need and have shown concern in the protection of intellectual property associated with biomaterials.

This international concern has resulted in several conferences, meetings and workshops that have eventually led to the production and signing of various conventions and agreements. Perhaps the most important and far-reaching of these efforts is the 'Biodiversity Convention.' The Convention on Biological Diversity, popularly known as the 'Biodiversity Convention,' came into force on December 29, 1993, having been drawn up and tabled at the United Nations in 1992. This Convention is a broad and legally binding writ that is probably the most important initiative yet taken to set the world on a course towards environmentally sustainable development. The Convention also supports natural sovereignty, the right of countries to benefit from their bioresources, and the right of countries to have access to technologies that could assist in the conservation and exploitation of their biological resources.

Various international conferences and workshops have also addressed the difficulties of patenting biological materials. In these various fora, many alternatives such as the option dubbed 'Fast GATT', the 'Patent Option,' the 'Union for the Protection of New Varieties of Plants (UPOV)', sui generis possibilities and the role of international biodiversity centers, have been formulated and extensively debated.20 However, despite the good intentions of the Biodiversity Convention, this covenant has several weaknesses that in the long run may make it difficult to implement in the interest of the less-industrialized countries. For example, genebank and botanic-garden materials collected before the coming into force of the Convention are unprotected by it, and so they are now beyond the reach of countries in the 'South' who initially were the major donors of these materials to conservation centers in the 'North'.

With the recent spate of attempts by various firms to patent biological materials originally collected from the 'South' and improved through genetic engineering, the 'South' may in the long run be the loser. In its report entitled "People, Plants and Patents," the Crucible Group had this to say about the Convention:

The Convention only applies to that material that we do not know to exist and that will probably not be commercialized in the foreseeable future. Unless otherwise established through agreed interpretations to the Convention, this new legal covenant, for the first time, acknowledges the right of governments and corporations that obtained the South's germplasm before the convention to declare this material their own and to control access to it and benefit from it. If this is the case, some members (of the Crucible Group) contend that the Biodiversity Convention of 1992 could become the biggest "rip-off" of indigenous peoples and of their knowledge and materials since 1492. Incidentally, 1492 marks the first major European settlement in Africa, through the construction of a castle in El Mina, Ghana (then The Gold Coast), by the Portuguese. This settlement marked the beginning of European trade in Africa, first in gold and ivory, then later—more ominously—in slaves. El Mina, and to a large extent Goree Island in Senegal, remained the biggest and most notorious slave trade center in the whole of Africa for a very long time. According to the Crucible Group, the range of feasible options for protecting the intellectual property of farmers and indigenous peoples has not yet been fully explored.

The Group has suggested that the FAO, UNESCO, UPOV, WIPO (and perhaps WHO) ought to be asked to convene an international meeting of experts to explore this issue in conjunction with industry, NGOs and farmers' organizations. Since whatever conclusions are derived can only be enforced through strong government backing, I believe that governments should also be involved in the organization of such meetings of experts.

It is a fact that none of the current Intellectual Property Conventions acknowledges fully the intellectual contribution of informal innovators such as herbalists and farmers. Hence, the incomparable expertise of these people—most of whom are citizens of 'developing' countries—is grossly undervalued. It is the absence of such recognition and acknowledgement that has led to the unquestioned and unchallenged appropriation of the innovations of rural communities.

INITIATIVES OF VARIOUS ORGANIZATIONS

In the absence of a comprehensive and fair policy on intellectual properties of indigenous peoples, various organizations have gone into bilateral contract arrangements with partners in the developing world, to take care of these shortcomings—despite the associated risks of such bilateral arrangements. The Merck/In Bio Initiative. In recognition of the value of biodiversity to industry, Merck (in collaboration with the NGO, In Bio) has set up a project in Costa

Rica, funding it with $US 1.135 million, for 10,000 extracts from medicinal plants collected by parataxonomists. They have also agreed on a royalty-sharing system if any of the materials are commercialized.

Various viewpoints have been expressed about this agreement. Some consider it as nothing but exploitation. They argue that whereas Merck's sales in 1991 were $US 8.6 billion, and their research budget was $1 billion, the total GNP of Costa Rica for that year was $5.2 billion—less than Merck's total sales. So what is a mere $US 1.135 million to Merck for 10,000 extracts? Others argue that this is an encouraging beginning in an area of partnership where nothing previously has existed. Yet another view is that for the exploitation of biodiversity, all options should be kept open to see which initiatives will stand the test of time.24 But can industrially developing countries afford such expensive experiments with their natural resource endowments and merely hope for some pay-off that may never materialize?

US National Cancer Institute/Missouri Botanical Gardens/University of Ghana. The National Cancer Institute and the Missouri Botanical Gardens, along with a number of African countries, have jointly established plant collection and screening programmes. The first was set up in Madagascar, followed by programmes in Cameroon and Tanzania. The latest is the one instituted about five years ago with the Botany and Chemistry Departments of the University of Ghana.

In the absence of adequate patent laws, the agreement for this particular project provides for:

- Capacity-building and strengthening in the research programmes of the University through staff exchange, material and equipment support;
- Royalty sharing if any material is commercialized.

  Through this project, the Missouri Botanical Gardens in collaboration with the Botany Department of the University of Ghana has collected a large number of plants from Ghana for screening by the National Cancer Institute (NCI) for potential anti-cancer and anti-AIDS drugs. There has been some staff exchange, and pure compounds isolated by the Chemistry Department have been tested using the facilities of the NCI. The three institutions are in constant discussion to improve the workings of the project within the framework of the original agreement. So far the only major area that has raised some concern is the capacity-building and institutional-strengthening aspect of the project with regard to the University of Ghana; this is currently under discussion.

The Shaman Partnership

Shaman Pharmaceuticals has announced its intention to return a percentage of profits back to all countries and communities that it has worked with after any and every product is commercialized. Compensation will be funneled through the Healing Forest Conservancy, a non-profit organization founded by Shaman for the conservation of biodiversity and for the protection of indigenous knowledge. Shaman's research has already led to patent claims, with full acceptance and recognition of the contribution of the communities from which it has received medicinal plants. The company has developed contracts with some indigenous communities in Latin America, but it will take some time before one can tell whether this arrangement has benefited the communities involved.

WIPO MODEL PROVISIONS FOR FOLKLORE

In 1985 The World Intellectual Property Organization (WIPO) and UNESCO published "Model Provision on Folklore." This model has three unique elements that seek to protect biological products and processes, including herbal preparations.

The provisions are as follows:

- Communities, rather than individuals, can be legally registered innovators and can either act on their behalf or be represented by the State.
- Community innovators are not necessarily fixed and finalized, but can be ongoing or evolutionary yet still protected by IP Law.
- Beyond standard patent or even copyright provisions, communities retain exclusive control over their folklore innovations for as long as the community continues to innovate.

One major weakness of these provisions is that they totally exclude scientific inventions. Therefore it is difficult to see how the IP of an indigenous people can be effectively protected under these provisions if a scientific invention, or discovery of a drug, or a new variety of a crop is made by a scientific laboratory.

After all, most local healers do not have the same knowledge as that of a chemist regarding the active compounds derived from a plant which are patented as medicine. So they will never know whether a particular drug put on the market emanated from germplasm collected from their own community or from folkloric information obtained from some member of their own community. A typical example is the current controversy raging over a compound, Michelamine—discovered in a West African plant Ancistrocladus corupensis, from the rain forest of Cameroon and believed to be one of the most encouraging discoveries yet made in the search for a cure for AIDS. But for the international hue and cry raised by other scientists from Africa, it is doubtful whether the indigenous peoples of Cameroon would ever have become aware of this discovery; indeed, I doubt that even now they are aware of it.

I have tried in this essay to examine various issues concerning the intellectual property rights of indigenous peoples of the less-industrialized world, especially as they apply to traditional medicines. The issues are even wider and more complex when it comes to food and agriculture. Biodiversity and conservation will certainly be one of the major issues of the twenty-first century, perhaps second only to information technology expansion and utilization. The time to examine these issues and protect the rights and interests of the less privileged societies who constitute the majority of the world's population is now.

GOLDEN SEAL CULTIVATION DESCRIPTION

It has been proven beyond all doubt that Golden Seal will not grow in the open field. That being the case we have two general lines of cultivation open for us to follow. One is to use natural shade, which comprises forest shade and also orchard and vine shading, and, in fact, any kind of shading where plants, shrubs, vines or trees are used to make the shade. The other method is by purely artificial means and using materials for the shade proper that do not draw either moisture or fertility from the soil. This consists of wood, metal and in some cases, vegetable fibre. The only reason Golden Seal will not grow in the open field is that the plant needs and must have shade of some kind. Were it not for the lack of shade, seal would grow in any of our fields where our common grains will grow. When we follow nature closely we are less liable to be troubled with disease such as blight and other fungi.

In the cultivations of this crop, we must bear in mind that nature in this case is very slow and if we follow nature we must expect also to go slow. I think, perhaps, it would be best to first consider the natural way of growing seal. Wild seal propagates in two ways, principally by the formation of new plants on the long slim fibre roots some distance away from the parent plant. The rhizome of the seal plant sends out, according to the vigour of the individual plant, from a half dozen up to as high as one hundred or more fibre roots, all nearly of one size and all very long and slim holding their full size to near the extreme end. Some of these roots are two feet in length and I have seen them over three feet.

When the root becomes fairly matured small plants start on the long slim roots and after two or three seasons, when this new plant gets well started, the root on which it formed will perish away somewhere between it and the parent plant, thus forming a separate and distinct plant. The other method of increase by the wild plant is by seed. The seed ripens the last of July or fore part of August and when first changing in colour from the green into a bright scarlet, and later turning to crimson, and if not disturbed or shaken from the plant, often dries and then turns to a very dark brown. If the seed remains dry for any length of time, it kills it outright. If by chance the seed falls soon after ripe and is soon covered by leaves so it will not dry out, it will grow the following spring. From close and long observation, I am inclined to think very few seeds ever grow in the wild state. Birds do scatter a few and they are more liable to grow for the reason that they are taken from the parent plant before drying and stripped of the pulp surrounding them. They are more liable to find their way when dropped among the leaves and litter to a place where they will not dry up.

This plant is purely a bedding plant and often is found as a thick compact mass of plants. It sometimes happens that they get so thick and crowded that rot sets in, usually where the roots are thickest. This starting in the center of a large old root rarely extends to the extreme ends and in that manner often makes several separate pieces of root with the center, and also the buds gone, but after a season or two, new buds will form on these pieces and thus practically new separate plants are formed. A plant from seed the first season in the forest is very small and does not have the proper seal leaf until the second year. The accompanying cut shows an average wild seedling the first season, both root and top.

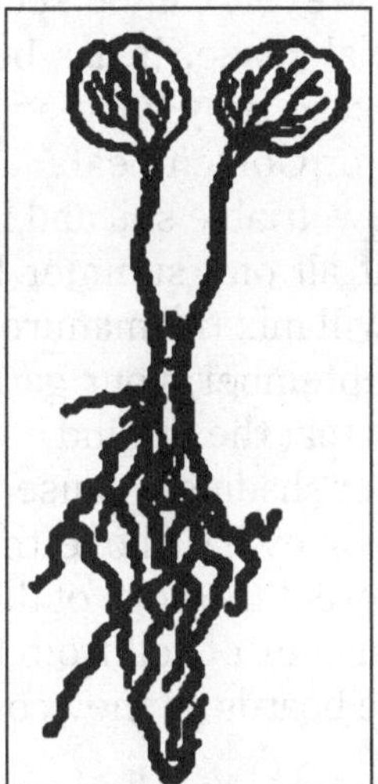

If the grower decides to select the forest for his garden, he should first remove all fibre root, of trees and large plants that come in the bed. He should also be very careful to select the location of each bed so as it will be well shaded during the middle of the day.

The large trees will throw their fibre roots into this mellow prepared bed and by the end of the second summer the gardener will find his beds crowded with them and the seal plants from that time on will make very slow growth, unless he again clears all these fibre roots out. To do this he will have to take up a few hundred first and heal them in some soft mellow soil where they will remain in good condition for some time without attention. Then clear out the roots and mellow up the plat of ground where he dug them and get it ready to plant again.

Then, as fast as he digs up other roots, set or plant them at once in the prepared bed. In this manner he can dig and reset his plant and not harm them by drying or exposing them to sun and wind. By transplanting them after two years, they will again make a good growth and by the end of the fourth summer will be ready to harvest. That is, if good, fair plants were used to start with.

A much larger growth, however, can be secured by growing this plant in good garden soil under artificial shade. It has been supposed that virgin soil must be used; this is not the case. Any good soil that will grow a fair crop of garden vegetables is all right for Golden Seal.

Select well drained, mellow, friable soil and if not rich, give it a heavy coat of stable manure, and work it all one summer by spading or ploughing and harrowing once a week. This will mix the manure and also sprout and get rid of weed seeds. By the first of September, your garden will be ready to plant. In preparing for planting, I advise that the ground be laid off into beds of a suitable width to accommodate whatever shading you use. I also find the plants to thrive better if the beds are raised somewhat above the level. We level the ground and then draw a line where we want the edge of the bed, and with round-pointed shovel throw out about eight inches of soil from the paths, placing this soil on the beds. I do not use the edge boards as they cost considerable and really are not needed.

In planting, we first determine how long we want the bed to stand before harvesting or resetting. Also much depends on the plants we have on hand. If we are planting good, strong plants and intend to dig them after three seasons' growth, we plant six by eight inches. Usually the rows eight inches apart and the plants six inches apart. If we desire them to remain four years, then we plant eight by eight inches, and if the bed is to remain more than four years undisturbed, they should be planted at least eight by ten inches. The reason for this is, that as this is a bedding plant and increases by adding new stalks from the rhizome and, also by new plants forming on the fibre roots, it, in time, becomes so thick that it would kill itself out if allowed to get too crowded. When it gets so thick that the entire ground is covered and the soil all taken up with little roots, it must be lifted. If not, the tops will decay close to the ground from lack of air and the roots themselves are very apt to decay. The crown or bud, in planting, should be placed about two inches under the surface and if the roots have many long, fibrous roots, we always, with a sharp knife, cut them off about two inches from the rhizome.

This helps in arranging the roots and spreading them out in planting, and also makes the plant fully as vigorous. The little ends, where cut, will usually throw out each two tiny plants soon after being put in the ground. Golden Seal is one of the few plants that can safely be planted any time of the year, when the ground is not frozen so it cannot be worked. During May and June is the time that most diggers of this wild root begin to collect it for planting. It is in blossom in this state in May. Some of the collectors of this root still leave the top on when they are collecting for planting in the garden, but this should not be done as the wilted top draws hard on the disturbed root. Take the top off as soon as dug and then if planted soon the roots will get hold and in about three weeks another top will appear and often this second top will bear seed that same season.

The gardener has several sources of supply to select from when buying his plants for starting a garden. First, because as yet the most abundant, comes the wild root as gathered by the root collectors. Then plants grown from seed in gardens, but in this connection the practice of planting yearling roots from seed is not to be advised, as they are too small to stand it. Two year old, grown in garden, is an ideal plant for transplanting. Another good plant comes from cutting up older garden plants and still another choice is the small plants started from layering pieces of fibre roots. We will describe each of this separately.

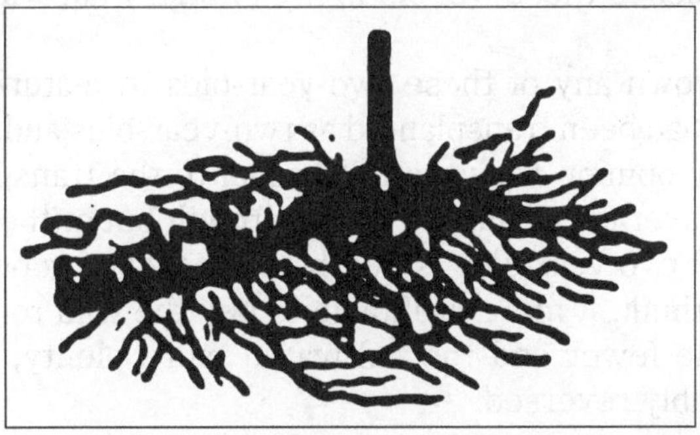

Fig. Is from a Photograph of a Wild Root but shows Rather more Fibres than the Average Wild Root as Received from the Collector.

It is a good root for planting and is about the size that you should expect when buying wild roots from a digger for your planting. Of course, some will be larger and some smaller but this is a fair average. Such roots as this should remain about four years after planting. They will bear to be planted about six by eight inches. If the ground is very rich and the plants do extra well, it may happen that at the end of the third summer after planting, that the roots will have fully occupied the ground and the tops also occupy all the overhead room. If this should be the case, you should harvest them, as another season's growth in, that crowded condition will surely invite decay and that will mean a heavy loss to the grower. Last fall (1913) the writer harvested a plot of this class of roots that had been planted three years and dried them and at market prices of dry root ($4.50). the roots brought some over $12,000.00 to the acre.

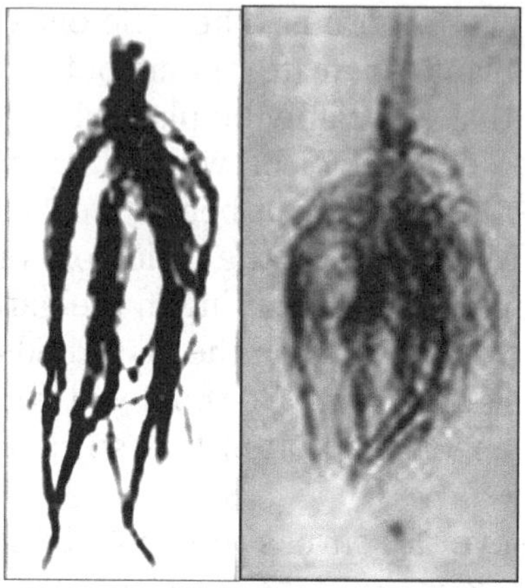

The next choice at the present time would be two-year roots from seed. See Fig. This is not a two-year from the forest or from a forest garden for, as a rule, such would not reach the size represented in this picture from a photo, which was three-fourths natural size. The coming plant will be these two-year-olds from gardens. At present there are not enough of them to supply the demand and, of course, the price is high. As soon as the cultivation of Golden Seal is well established these will be, in my I mind, what will be sought for by the planters.

I have not grown any of these two-year-olds to maturity but have seen them where they had been transplanted as two-year-olds and then grown three years and it is my opinion they are fully equal to the transplanted wild roots and probably will average to make a larger growth but at the present time the price of cultivated two year-olds is nearly twice that of good wild roots. With that difference, I think, it more profitable to buy the wild root for planting. As wild roots become fewer and the cultivated more plenty, the price will be changed and possibly reversed.

Another grade of roots for planting is shown in Fig. These plants are divisions of large plants. The reset wild plants make better root to divide than the strictly cultivated plants, as the rhizome is larger in proportion to the fibre roots. The plants represented in Fig. were taken from mature roots by cutting into lengths of about one-half inch of the rhizome and the cutting was so planned as to give each piece a bud. Parts of the root where a bud could not be given may be dried for market or if plants are greatly to be desired these may be planted and they will form buds but will be about one year behind those that had good strong buds. While most of them will form buds and come up the first season, they will not be as strong and growthy as the others. To make this class of plants one need not wait for maturity of root but may dig up a plot that has had but one season's growth and they will be able to make at least two plants from one on an average. This class of plants are stronger than either of the two first named, but considering their value to dry and sell as dry root, they are really more costly than the others. However, I am planting these in preference to any others, when I can get them.

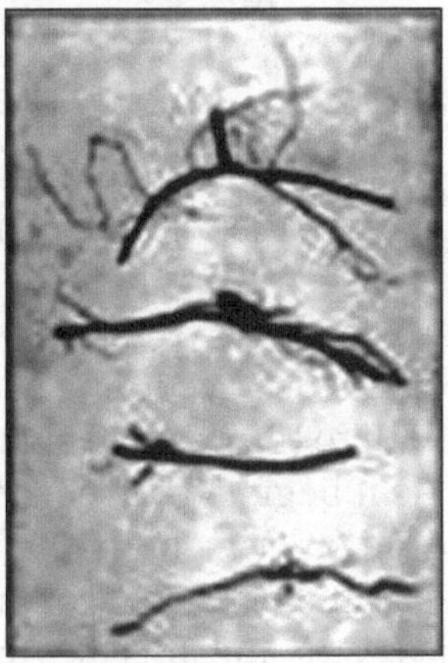

Still another class of plants for the gardener to consider are those represented in Fig. The four little plants shown in the photo at Fig. were produced by layering pieces of fibre roots in soft loam with a little gray sand added, but are identical with the little plants formed on the fibre roots of mature plants and in harvesting a crop of four-year-olds for drying, enough of these would be found to reset the plot.

If the gardener is anxious to increase his garden he can, from one mature plant, cut from one hundred to five hundred of these pieces and by layering get from fifty to seventy-five per cent of them to form buds. The two upper ones in the cut have had one season's growth and another year they will grow rapidly. The two lower ones were cut from the parent plant ten weeks before the picture was taken and at the' time of layering showed no indications of forming a bud. At that time, out of two thousand pieces layered in November and placed in a warm cellar as an experiment, about two hundred had formed buds.

It is, at this writing, about three weeks since we counted the two hundred that had formed buds, and today we find forty-five more. This process will continue for nearly a year but nearly all will form buds sooner or later. This class of plants grow slow the first year but after they are established, make good, thrifty stock. This is the favourite manner of increase of this plant in its wild state.

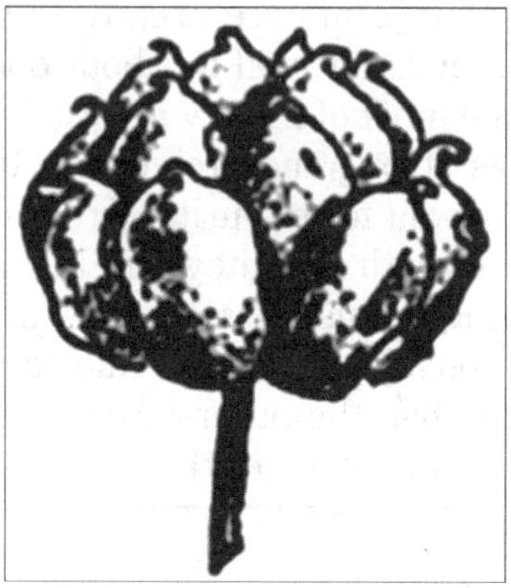

Fig. Caring for the Seed.

Golden Seal Seed is more difficult to handle than most seeds the gardener is accustomed to and the ordinary farmer or city man is doomed to failure unless he carefully posts himself, First, get it firmly fixed in the mind that at no time should this seed be allowed to get dry clear through.

The seed ripens in July or August, according to the locality where it grows. The seed head contains from ten to thirty small shiny black seeds and is similar to a common red raspberry. As it begins to turn from green to red the colour is scarlet and as it begins to verge on crimson is the best time to gather.

As soon as gathered, separate the seeds from the pulp. This may be done by placing the berries in a sieve, the meshes of which will not allow the seeds to pass through and then rub and crush with the hand; then with a copious supply of water wash out all the juice and pulp possible.

By persistent rubbing and washing most of the juice and pulp can be eliminated; then layer the seeds with at least twice their bulk of fine sifted loam and let stand a month, when the seed will be ready to plant.

Another method of getting rid of the pulp is to run the berries through an Enterprise meat chopper, being careful to use such attachments as will not crush very many seeds. Then wash and treat as above.

This seed grows the spring following the harvest and should always be planted in the fall. If the attempt is made to carry it over until warm weather comes in spring, it is almost sure to sprout and spoil.

It can be kept over, however, by placing it layered in the loam in boxes with perforated tops and bottoms and burying the box in the garden, where it will be kept cold and even freeze. Fig. Shows the seed or upper leaf on the stalk and also the seed berry.

HARVESTING, DRYING AND MARKETING

The digging of Golden Seal should not be undertaken when the soil is very dry, especially if soil is heavy, as many of the fibre roots will break off and be lost. The gardener should also wait until the leaves and tops have died down, as that is the time when they weigh the heaviest and are the strongest in medicinal properties.

If the tops have been killed with frost, wait two weeks to let them harden and mature, as they were-killed before this process was completed. A spading fork is, a good thing.

To dig with; shake off what dirt you can as soon as dug as it comes off better then. In washing, great care should be taken to get them clean. An old splint basket or large sieve with a good stream of water running into it is all right. Very large roots will have to be broken apart in order to get the dirt clinging to the rhizome among the thick sprangles of fibre roots. This breaking and dividing does not hurt their market value as it does in the case of Ginseng.

If there are small roots to be saved for replanting, this should be done as soon as the roots are dug and the little fellows carefully healed in the garden or moist loam in boxes. The drying is very simple. Spread thin on floor of a warm room or place on screens over moderate heat. Be sure that the larger parts of the root are perfectly dry before putting away, as it will cause the whole to mould if these are still damp.

The drying should be done in the shade. This root needs no sorting or other preparations for market after it is dry. Select a clean, light box to pack it in, line the box with paper, put in a few inches of roots and fibres, just as they grew, and then press down firmly.

I do this by setting box on the floor and using my feet, placing my whole weight on them. This breaks many of the fibre roots but it does no harm.

Wild roots are worth the most because they have the least fibre in proportion to the weight of rhizome. In other words, the rhizome is the strongest part of the root and the price is based on the proportion of that part of the root as compared with the fibre roots.

General Remarks

That the general reader may get an idea of the difference in growth of the cultivated root and the wild, we give two illustrations, life size.

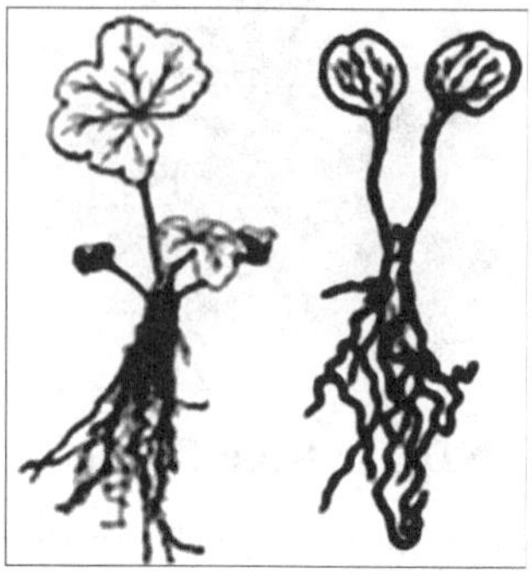

Fig. is a yearling, cultivated from seed, and has had one season's growth. Fig. is precisely the same age but is a wild, or forest grown, plant. Note the difference in the leaf; the wild seedling has not as yet developed the true seal leaf and has only the two little round baby leaves, while the garden plant has two distinct seal leaves.

This marked difference in growth will follow throughout the life of the plant unless great care is taken to keep the roots of the trees and larger forest plants from robbing the seal of moisture and fertility. To give some idea as to what can be done we have known cultivated seal roots to reach fourteen, sixteen, and in one case, eighteen ounces, dry weight, and this in only six years, from the wild plant and in a country where seal does not grow wild.

The above large grow th occurred on the Pacific Coast. The largest we have seen grown in the north was one pound, green weight, at five years. The seal grower, however, should be cautioned against high fertilizing, as it is liable to cause disease. Acid Phosphate should not be used, as we have known several cases where it caused the immediate death of the plants. Neither should raw manure be used. Well rotted stable manure is the best fertilizer we know of and is also best mulch. It is not necessary to use mulch but in many cases it is advisable as it conserves moisture and keeps back many small weeds. Our own gardens of seal we cultivate the top about one and one-half inches deep and keep it very fine and mellow all summer.

This acts as a mulch and keeps the soil always moist and the plants growing. Care should be used in mulching seed beds as too heavy a mulch will cause a vacant spot in your beds the next summer. One inch of saw dust or buckwheat hulls is the limit and one and one-half inches of well rotted manure. Leaves, unless well rotted, should never be used on seed beds unless taken off early in the spring. Golden Seal is unlike Ginseng, in that the market for it is based on its real value. From this plant are extracted the alkaloids, Hydrastinine and also Berberine, two preparations used by every physician in the country. Besides there is a large export demand. The growing of this plant must be resorted to if the drug trade is to be supplied with these alkaloids, and it is opening a very large and, profitable field.

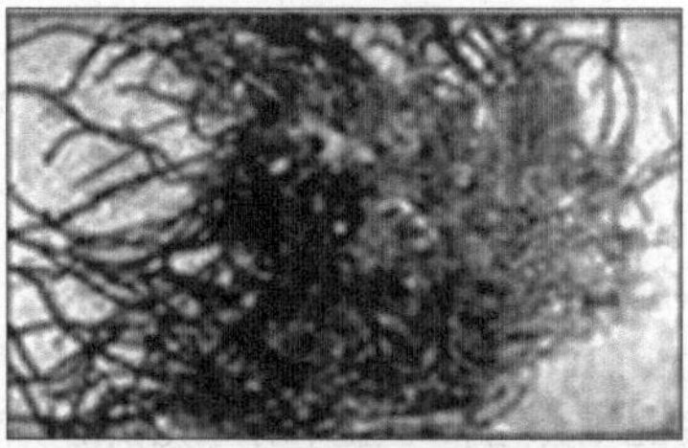

The cultivated root, when it attains age, becomes an irregular shaped chunked root or mass of roots. The first season after planting it usually throws up but one stalk but during the summer one or more new buds will start from other parts of the rhizome and the second summer two to four stalks will appear and this process continues until we find an old plant with as high as twenty stalks and as the root enlarges where each stalk comes out, we have a very irregular mass not at all resembling the one straight, little, wild root planted. Figure is a medium-sized root of this sort, weighing, green, six ounces.

This far no disease has attacked the seal plant, and if growers do not manure it too highly, we are not looking for any, but we should all remember that for ages and ages, this plant has been pinched for food and has been taught to rough it. Slow growth in nature means stability and if we reverse the conditions and force the plant by coddling and high feeding, we are liable to undermine the constitution, no matter how rugged. In shading this plant, the general instructions given for Ginseng will suffice and I should add here that the more sun you give it, up to the point of burning, the larger root growth you will get and not only that but also the larger per cent of Hydrastinine, or Hydrastine, as it is usually spelled. The U. S. Official Pharmacopia calls for this roof to analyse two and five-tenths per cent of Hydrastine.

The wild root, dug in the fall, will exceed this by four or fivetenths of one per cent, sometimes reaching three per cent of Hydrastine. Spring dug root, although wild, will barely reach the standard. Cultivated root, fall dug, rhizome and fibres as they grew, will exceed the standard, but the fibre roots alone will not come up to the standard. This root loses somewhat of its medicinal value if kept any great length of time, after it is dried. The rhizome, or root proper, and also the fibre roots, are the parts used in medicine.

Of late years the tops of this plant have come into use and now command about the same price that the roots did twenty years ago, which is around twenty cents per pound. Since 1894 the price has advanced from about twenty cents per pound until in 1911 the price reached $4.50 and for extra quality, a little above that. In 1912, owing to the high price of the year before more than usual was marketed and the price dropped for a time but recovered to $4.50 or better in 1913. To properly plant and shade one acre, with good plants and good substantial lumber shade, cost at present a little over $2,000.00 This will vary some in different localities, as the price of roots and lumber and labour will vary, but will come close to this, estimate.

The returns from one acre, at four years from planting, will average $10,000.00 or better and at five years would reach an average of at least $15,000.00 and in many cases will go far beyond this and, of course, with the careless and slipshod grower will fall far short of it. learned when a boy, by actual experience, that Golden Seal and Ginseng will not grow in open cultivated fields or gardens, I tried it faithfully. The soil must be virgin, or made practically so by the application of actual "new land" in such quantities that to prepare an acre for the proper growth of these plants would be almost impossible. And to furnish and keep in repair artificial shade for, say, an acre, would cost quite a little fortune. Of course one may cultivate a few hundred or few thousand in artificially prepared beds and shaded by artificial means, but to raise these plants successfully in anything like large quantities we must let nature herself prepare the beds and the shade.

When we follow nature closely we will not be troubled with diseases, such as blight and fungus. I know this by actual experience dear, and therefore dear to me. Plants propagate themselves naturally by seedage, root suckers, and by root formation upon the tips of pendulous boughs coming in contact with the ground. Man propagates them artificially in various ways, as by layering, cuttings, grafting or budding, in all of which he must follow nature.

The Golden Seal plant is readily propagated by any of the three following methods:

1. By seed;
2. By division of the large roots;
3. By suckers, or by small plants which form on the large fibrous roots.

The seed berries should be gathered as soon as ripe, and mashed into a pulp, and left alone a day or two in a vessel, then washed out carefully and the seed stored in boxes of sandy loam on layers of rock moss, the moss turned bottom side up and the seed scattered thickly over it, then cover with about one-half inch of sandy loam, then place another layer of moss and seed, until you have four or five layers in a box. The box may be of any convenient size.

The bottom of the box should be perforated with auger holes to secure good drainage. If water be allowed to stand upon the seeds they will not germinate, neither will they germinate if they become dry. The seeds should be kept moist but not wet. They may be sown in the fall, but, I think the better way, by far, is to keep your box of seeds in a cellar where they will not freeze until the latter part of winter or very early spring. If your seeds have been properly stratified and properly kept you will find by the middle of January that each little black seed has burst open and is wearing a beautiful shining golden vest. In fact, it is beginning to germinate, and the sooner it is put into the seed-bed the better. If left too long in the box you will find, to your displeasure, a mass of tangled golden thread-like rootlets and leaflets, a total loss.

To prepare a seed-bed, simply rake off the forest leaves from a spot of ground where the soil is rich and loamy, then with your rake make a shallow bed, scatter the seeds over it, broadcast, being careful not to sow them too thick. Firm the earth upon them with the back of the hoe or tramp them with the feet. This bed should not be near a large tree of any kind, and should be protected from the sun, especially from noon to 3 P.M.

The Golden Seal seedling has two round seed leaves upon long stems during the first season of its growth. These seed leaves do not resemble the leaves of tile Golden Seal plant. The second and usually the third years the plant has one leaf. These seedlings may. be set in rows in beds for cultivation in the early spring of the second or third year. This plant grows very slowly from seed for the first two or three years, after which the growth is more satisfactory.

By the second method, i. e., by division of the large roots, simply cut the roots up into pieces about one-fourth inch long and stratify in the same way as recommended for seeds, and by spring each piece will have developed a bud, and will be ready to transplant into beds for cultivation. This is a very satisfactory and a very successful method of propagating this plant. The plants grow off strong and robust from the start and soon become seed bearing.

By the third method we simply let nature do the work. If the plants are growing in rich, loose, loamy soil, so the fibrous roots may easily run in every direction, the whole bed will soon be thickly set with plants. These may be taken up and transplanted or may be allowed to grow and develop where they are. This is the method by which I propagate nearly all of my plants. It is a natural way and the easiest of the three ways to practice. As to the proper soil and location for a Golden Seal garden I would recommend a northern or northeastern exposure. The soil should be well drained and capable of a thrifty growth of deciduous trees. It should contain an ample supply of humus made of leaf mold. It will then be naturally loose and adapted to the growth of Golden Seal. Cut out all undergrowth and leave for shade trees that will grow into value. I am growing locust trees for posts in my Golden Seal garden. I do not think fruit trees of any kind suitable for this purpose. In preparing the ground for planting simply dig a trench with a mattock where you intend to set a row. This loosens up the soil and makes the setting easy. Set the plants in this row four to six inches apart. For convenience I make the rows up and down the hill. In setting spread the fibrous roots out each way from the large main root and cover with loose soil about one to two inches deep, firming the soil around the plant with the hands. Be very careful not to put the fibrous roots in a wad down in a hole.

They do not grow that way. Plants may be set any time through the summer, spring or fall, if the weather be not too dry. The tops will sometimes die down, in which case the root will generally send up a new top in a few days. If it does not it will form a bud and prepare for growth the next spring. The root seldom if ever dies from transplanting. I know of no plant that is surer to grow when transplanted that Golden Seal. I make the rows one foot to fifteen inches apart. It does not matter as it will soon fill the spaces with sucker plants any way.

The cultivation of Golden Seal is very simple. If you have a deep, loose soil filled with the necessary humus your work will be to rid the plot of weeds, and each fall add to the fall of forest leaves a mulch of rotten leaves. Do not set the plants deeper than they grew in a natural state, say about 1/2 to 3/4 inch. Spread the fibrous roots out in all directions and cover with leaf mold or some fine, loamy new soil. Water if the ground be at all dry. Then mulch with old forest leaves that have begun to decay. Let the mulch be about three or four inches deep and held on by a few light brush. The wind would blow the leaves away if not thus held in place. Be careful, however, not to press the leaves down with weights. Remove the brush in the early spring, but let the leaves remain. The plants will come up thru them all right. This plants grows best in a soil made tip entirely of decayed vegetation, such as old leaf beds and where old logs have rotted and fallen back to earth.

If weeds or grass begin to grow in your beds pull them up before they get a start. Be careful to do this. Do not hoe or dig up the soil any way. The fibrous roots spread out in all directions just under the mulch. To dig this up would very much injure the plants. I think the plants should be set in rows about one foot apart, and the plants three or four inches apart in the rows. This would require about 1,000 plants to set one square rod. My Golden Seal garden is in a grove of young locust trees that are rapidly growing into posts and cash. The leaves drop down upon my Golden Seal and mulch it sufficiently. The locust belongs to the Leguminous family of plants, so while the leaves furnish the necessary shade they drink in the nitrogen from the atmosphere and deposit great stores of it in the soil. This makes the soil porous and loose and gives the plant a very healthy dark green appearance.

We have only to follow the natural manner of the growth of the Golden Seal to be successful in its culture. Select a piece of sloping land, so as to be well drained, on the north or northeast side of the hill—virgin soil if possible. Let the soil be rich and loamy, full of leaf mold and covered with rotting leaves and vegetation. This is the sort of soil that Golden Seal grows in, naturally. It is hard to fix up a piece of ground, artificially, as nature prepares it, for a wild plant to grow in. So select a piece if possible that nature has prepared for you. Do not clear your land. Only cut away the larger timber. Leave the smaller stuff to grow and shade your plants. There is no shade that will equal a natural one for Ginseng or Golden Seal. Now, take a garden line and stretch it up and down the hill the distance you want your bed to be wide. Mark the place for the row along the line with a mattock, and dig up the soil to loosen it, so as to set the plants, or, rather, plant the roots easily. With a garden dibble, or some other like tool make a place for each plant. Set the plants 4 to 6 inches apart in the row.

The crown of the plant or bud should be set about I inch beneath the surface. Firm the earth around the plant carefully. This is an important point and should be observed in setting any plant. More plants are lost each year by carelessly leaving the earth loose over and around the roots than from any other cause. Do not leave a trench in the row. This may start a wash. Let the rows be about I foot apart. If land is no item to you, the rows may be further apart. They will, if properly cared for, in a few years, by sending up sprouts from the roots, fill up the end completely.

When you have finished setting your bed, cover it with a good mulch of rotten leaves from the forest and throw upon them some brush to keep the wind from blowing them away. By spring the leaves will settle down compactly and you will be pleased to see your plants grow luxuriously. October and November are the best months of the year in which to set Golden Seal plants. They are, also, the months in which it should be dug for the market. It may be set in the spring if the plants are near by. The roots will always grow if not allowed to dry before transplanting. If your bed does not supply you with plants fast enough by suckering, you may propagate plants by cutting the roots into pieces about one-fourth inch long, leaving as many fibrous roots on each piece as possible. These cuttings should be made in September or October and placed in boxes of sand over winter.

The boxes should be kept in a cellar where they will not freeze. By spring these pieces will have developed a bud and be ready for transplanting, which should be done just as early as the frost leaves the ground so it can be worked. All the culture needed by this plant is to mulch the beds with forest leaves each fall and keep it clear of grass and field weeds. Wild weeds do not seem to injure it. Golden Seal transplants easily and responds readily to proper cultivation. There is no witchcraft in it. The seeds ripen in a large red berry in July to germinate, if planted at once, the next spring. The fibrous roots, if stratified in sand loam in the autumn, will produce fine plants. Any good, fresh, loamy soil, that is partially shaded will produce a good Golden Seal.

You want soil that is in good tilth, full of humus and life, and free from grasses and weeds. It will stand a great deal more sunlight than Ginseng. It will also produce a crop of marketable roots much quicker than Ginseng. There is no danger of an over supplied market, as the whims of a nation changing, or of a boycott of a jealous people. I have my little patch of Golden Seal that I am watching and with which I am experimenting. I want to say right here that you do not need a large capital to begin the culture of these plants that are today being exploited by different parties for cultivation. just get a little plot of virgin soil, say six yards long by one yard wide and divide it into two equal lots.

Then secure from the woods or from some one who has stock to sell about 100 plants of each, then cultivate or care for your apron garden and increase your plantation from your beds as you increase in wisdom and in the knowledge of the culture of these plants. The Bible says "Despise not the day of small things." Do not, for your own sake, invest a lot of money in a "Seng" or Seal plantation or take stock in any exploiter's scheme to get rich quick by the culture of these plants. Some one has written a book entitled "Farming by Inches." It is a good book and should be in every gardener's library. Now, if there be any crops that will pay a big dividend on the investment farmed by inches, "Seng" and Seal are the crops.

GOLDEN SEAL HISTORY

The increasing use of Golden Seal in medicine has resulted in a wide demand for information about the plant, its identification, geographical distribution, the conditions under which it grows, methods of collecting and preparing the rhizomes, relations of supply and demand, and the possibilities of its cultivation. This paper with the exception of the part relating to cultivation wa s prepared (under the direction of Dr. Rodney H. True, Physiologist in Charge of Drug and Medicinal Plant Investigations) by Miss Alice Hinckel, Assistant in Drug and Medicinal Plant Investigations; and Mr. G. Fred Klugh, Scientific Assistant in the same office, in charge of Cultural Experiments in the Testing Gardens, furnished the part treating of the cultivation of this plant. In the preparation of this paper, which was undertaken to meet the demand for information relative to Golden Seal, now fast disappearing from our forests, many facts have been obtained from Lloyd's Drugs and Medicines of North America.

HISTORY

As in the case of many other native medicinal plants, the early settlers learned of the virtues of Golden Seal thru the American Indians, who used the root as a medicine and the yellow juice as a stain for their faces and a dye for their clothing. The Indians regarded Golden Seal as a specific for sore and inflamed eyes and it was a very popular remedy with pioneers of Ohio and Kentucky for this affection, as also for sore mouth, the root being chewed for the relief of the last named trouble. Barton in his "Collection for an Essay towards a Materia Medica of the United States," 1804, speaks of the use of a spiritual infusion of the root of Golden Seal as a tonic bitters in the western part of Pennsylvania and the employment of an infusion of the root in cold water as a wash for inflammation of the eyes.

According to Dr. C. S. Rafinesque, in his Medical Flora in 1829, the Indians also employed the juice or infusion for many "external complaints, as a topic tonic" and that "some Indians employ it as a diuretic stimulant and escharotic, using the powder for blistering and the infusion for the dropsy." He states further that "internally it is used as a bitter tonic, in infusion or tincture, in disorders of the stomach, the liver," etc. It was not until the demand was created for Golden Seal by the eclectic school of practitioners, about 1747 (actually 1847), that it became an article of commerce, and in 1860 the root was made official in the Pharmacopoeia of the United States, which place it has held to the present time.

HABITAT AND RANGE

Golden Seal occurs in patches in high open woods where there is plenty of leaf mold, and usually on hillsides or bluffs affording nature drainage, but it is not found in very moist or swampy situations, in prairie land, or in sterile soil. It is native from southern New York to Minnesota and western Ontario, south to Georgia and Missouri, ascending to an altitude of 2,500 feet in Virginia. It is now becoming scarce thruout its range.

Not all of this region, however, produced Golden Seal in. abundance. Ohio, Indiana, Kentucky and West Virginia have been the greatest Golden Seal producing states, while in some localities in southern Illinois, southern Missouri, northern Arkansas, and central and western Tennessee the plant, tho common, could not be said to be sufficiently plentiful to furnish any large amount of the root. In other portions of its range it is sparingly distributed.

COMMON NAMES

Many common names have been applied to this plant in different localities, most of them bearing some reference to the characteristic yellow colour of the root, such as yellow root, yellow puccoon, orange-root, yellow paint, yellow Indian paint, golden root, Indian dye, curcuma, wild curcuma, wild turmeric, Indian turmeric, jaundice root and yellow eye; other names are eyebalm, eye-root and ground raspberry.

Yellow root, a popular name for it, is misleading, as it has been applied to other plants also, namely, to gold thread, false bittersweet, twinleaf and the yellow-wood. The name Golden Seal, derived from its yellow colour and seal-like scars on the root, has been, however, generally adopted.

DESCRIPTION OF THE PLANT

It is a perennial plant and the thick yellow rootstock sends up an erect, hairy stem about a foot in height, around the base of which are two or three yellowish scales. The stems, as they emerge from the ground, are bent over, the tops still remaining underground, and sometimes the stems show some distance above the surface before the tops are brought out from the soil. The yellow colour of the roots and scales extends partly up the stem so far as it is covered by soil, while the portion of the stem above the ground has a purplish colour. Golden seal has only two leaves (rarely three), the stem bearing these seeming to fork at the top, one branch supporting a large leaf and the other a smaller one and a flower. Occasionally there is a third leaf, much smaller than the other two and stemless.

The leaves are prominently veined on the lower surface, and are palmately 5 to 9 lobed, the lobes broad, acute, sharply and unequally toothed. The leaves are only partially developed at flowering time and are very much wrinkled, but they continue to expand until they are from 6 to 8 inches in diameter, becoming thinner in texture and smoother. The upper leaf subtends or encloses the flower bud. Early in spring, about April or May, the flower appears, but few ever see it as it lasts only five or six days.

It is greenish-white, less than half an inch in diameter, and has no petals, but instead three small petal-like sepals, which fall away as soon as the flower expands, leaving only the stamens-as many as 40 or 50—in the center of which are about a dozen pistils, which finally develop into a round, fleshy, berry-like head. The fruit ripens in July or August, turning a bright red and resembling a large raspberry, whence the common name ground raspberry, is derived.

Each fruit contains from 10 to 20 small, black, shining, hard seeds. If the season has been moist, the plant sometimes persists to the beginning of winter, but if it has been a dry season it dies soon after the fruit is ripe, so that by the end of September no trace of the plant remains above the ground. In a patch of Golden Seal there are always many sterile stems, simple and erect, bearing a solitary leaf at the apex but no flower.

Mr. Homer Bowers, of Montgomery county, Ind., who propagated Golden Seal from the seed for the purpose of studying its germination and growth, states that the plant grown from naturally sown seed often escapes observation during the first year of its existence owing to the fact that in this entire period nothing but two round seed leaves are produced and at this stage the plant does not look materially different from other young seedlings. During its second year from seed one basal leaf is sent up, followed in the third year by another smaller leaf and the flower.

DESCRIPTION OF THE RHIZOME, OR ROOTSTOCK

The rhizome (rootstock) and rootlets of Golden Seal, or hydrastis, as it is also known in the drug trade, are the parts employed in medicine. The full-grown rhizome, when fresh, is of a bright yellow colour, both internally and externally, about 1 1/2 to 2 1/2 inches in length, and from one-fourth to three-fourths of an inch in thickness. Fibrous yellow rootlets are produced from the sides of the rhizome. The fresh rhizome contains a large amount of yellow juice, and gives off a rank, nauseating odour. When dry the rhizome measures from one to two inches in length and from one-eighth to one-third of an inch in diameter. It is crooked, knotty, wrinkled, of a dull brown colour outside, and breaks with a clean, short, resinous fracture, showing a lemon-yellow colour if the root is not old. If the dried root is kept for a long time it will be greenish-yellow or brown internally, and becomes inferior in quality. On the upper surface of the rhizome are several depressions, left by former annual stems, which resemble the imprint of a seal; hence the name Golden Seal. The fibrous rootlets become very wiry and brittle in drying, break off readily and leaving only small protuberances, so that the root as found in commerce is sometimes almost bare. The dried rhizome has also a peculiar, somewhat narcotic, disagreeable odour, but not so pronounced as in the fresh material; an exceedingly bitter taste; and a persistent acridity which causes an abundant flow of saliva when the rhizome is chewed.

COLLECTION AND PREPARATION OF THE ROOT

The root should be collected in autumn after the plants have matured. Spring-dug root shrinks far more in drying and always commands a lower price than the fall-dug root. After the roots are removed from the earth they should be carefully freed from soil and all foreign particles. They should then be sorted and small, undeveloped roots and broken pieces may be laid aside for replanting. After the roots have been cleaned and sorted they are ready to be dried or cured.

Great care and judgment are necessary in drying the roots. It is absolutely necessary that they should be perfectly dry before packing and storing, as the presence of moisture induces the development of molds and mildews, and of course renders them worthless. The roots are dried by the exposure to the air, being spread out in thin layers on drying frames or upon a large, clean, dry floor.

They should be turned several times during the day, repeating this day after day until the roots are thoroughly dried. If dried out of doors they should be placed under cover upon indication of rain and at night so that they may not be injured by dew. After the roots are thoroughly dried they may be packed as tightly as possible in dry sacks or barrels and they are then ready for shipment.

DIMINUTION OF SUPPLY

Altho, perhaps, in some secluded localities Golden Seal may still be found rather abundantly, the supply is rapidly diminishing and there is a growing scarcity of the plant thruout its range. With the advance of civilization and increase in population came a growing demand for many of our native medicinal plants and a corresponding decrease in the sources of supply. As the rich forest lands of the Ohio valley and elsewhere were required for the needs of the early settlers they were cleared of timber and cultivated, and the Golden Seal, deprived of the shelter and protection necessary to its existence, gradually disappeared, as it will not thrive on land that is cultivated.

Where it was not destroyed in this manner the root diggers, diligently plying their vocation, did their share towards exterminating this useful little plant, which they collected regardless of the season, either before the plants had made much growth in the spring or before the seeds had matured and been disseminated, thus destroying all means of propagation. The demand for the root appears to be increasing, and the time seems to be not far distant when this plant will have become practically exterminated, so far as the drug supply is concerned.

The cultivation of golden seal seems now to have become a necessity in order to meet the demand and save the plant from extinction. Prior to 1900 there seemed to be no one, so far as the Department of Agriculture could ascertain, who had ever attempted the cultivation of golden seal for the market. From that time on, many inquiries were directed to the Department by persons who were quick to note the upward tendency of prices for golden seal and there are now several growers in different parts of the country who have undertaken the cultivation of golden seal on a commercial scale.

CULTIVATION

The United States Department of Agriculture has been carrying on experiments in the cultivation of Golden Seal on a small scale at Washington, D. C., since the spring of 1899, in the hope that methods might be worked out according to which this valuable wild drug plant could be grown on a commercial scale. In these experiments the aim has been to imitate the natural conditions of growth as closely as possible. The results that have thus far been obtained, while not as complete in some respects as would be desirable, seem to justify the conclusion that Golden Seal can be successfully cultivated. The methods of operation described apply to the conditions at Washington, and the treatment may need to be somewhat modified under other conditions of soil and climate.

NECESSARY SOIL CONDITIONS

The soil conditions should imitate as closely as possible those seen in thrifty deciduous forest. The soil should contain an ample supply of humus, well worked into the ground, to secure the lightness and moisture-retaining property of forest soils. The best form of humus is probably leaf mold, but good results may be obtained by mulching in the autumn or early winter with leaves, straw, stable manure, or similar materials.

After the soil has been prepared and planted, it is well to add a mulch in the fall as a partial protection to the roots during the winter, and the decay of this material adds to the value of the soil by the time the plants appear in the spring. The forest conditions are thus imitated by the annual addition of vegetable matter to the soil, which by its gradual decay accumulates an increasing depth of a soil rich in materials adapted to the feeding of the plants and to the preservation of proper physical conditions.

The growth of the weeds is also hindered to a considerable extent. If sufficient attention is given to the presence of this mulch, the nature of the underlying soil is of less importance than otherwise. In the case of clay the thorough incorporation of a large amount of decayed vegetable matter tends to give lightness to the otherwise heavy soil, facilitating aeration and drainage. Since the roots of the Golden Seal do not grow well in a wet soil, thorough drainage is necessary. A lighter, sandy soil is improved by the addition of humus, since its capacity to hold moisture is thereby increased and the degree of fertility is improved. The looser the soil, the easier it is to remove the roots in digging without breaking or injuring them. Before planting, the soil should be thoroughly prepared to a depth of at least 6 or 8 inches, so as to secure good aeration and drainage. The good tilth thus secured will be in a degree preserved by the continued addition of the mulch. A further advantage of a careful preparation is seen in a decrease in the amount of cultivation required later.

ARTIFICIAL SHADE

Since the Golden Seal grows naturally in the woods, it must be protected from the full light of the sun by artificial shade. That used in connection with the experiments of the Department was made of ordinary pine plastering lath, nailed to a suitable frame elevated on posts. The posts were of cedar 81 feet long, set 21 feet in the ground in rows 11 feet apart, and 16 feet distance from each other in the rows. Supports 2 by 4 inches were set on cedar blocks 2 feet long -sunk below the soil surface in the middle of the 16-foot spaces. Pine pieces 2 by 4 inches were nailed edgewise to the tops of the posts and supports.

The posts were notched to receive the 2 by 4-inch sticks. Pieces 2 by 4 inches were nailed across these at intervals of 4 feet. The laths were nailed to these, leaving spaces about an inch wide. This shade has been found to be satisfactory, as it is high enough above the ground to allow such work as is necessary in preparing and cultivating the land.

If the lathing is extended 2 or 3 feet beyond the posts on the sunny sides, injury from the sun's rays at the edges of the area will be prevented. The sides may be protected by portable board walls about 2 feet high set around the edges. Protection from injury by winds when the tops are large may be thus secured. Too much dampness should be guarded against in the use of the board sides, since conditions might be developed favourable to the damping off fungus and to aphides during the hot, rainy periods. Trees may be used for shade, but this is in some ways to be regarded as unsatisfactory. When the shade produced is of the right density, the use of the moisture and raw food materials of the soil by the trees is an undesirable feature.

ATTENTION REQUIRED

The cultivation of Golden Seat is simple. Having secured a deep, loose soil, rich in humus, renewed annually by the application of a new mulch, the removal of weeds is the chief care. The soil, if properly prepared, will tend to maintain itself in good condition. The manner of treatment is very similar to that required by Ginseng, which is also a plant of moist woods. If the ground is thoroughly prepared, beds are not absolutely necessary. The plants may be grown in rows 1 foot apart and 6 inches apart in the rows. Beds may be thought by some to be more convenient, enabling the grower to remove the weeds and collect the seed more readily. If beds are used, they may be made from 4 to 8 feet wide, running the entire length of the shade, with walks from 18 inches to 2 feet wide between. Boards 6 or 8 inches wide are set up around the sides of the beds, being held in place by stakes driven on each side of the board in the center and at the ends. These beds are filled with prepared soil, and the plants are set 8 inches apart each way.

METHODS OF PROPAGATION

There are three possible ways of propagating the plant:

1. By seed;
2. By division of the rhizomes;
3. By means of small plants formed on the stronger fibrous roots. Thus far no success has been attained in growing Golden Seal from the seed. The second and third methods have given better results.

EXPERIMENTS WITH SEEDS

Seeds just after ripening were planted in sandy soil mixed with well rotted stable manure and mulched lightly with manure. Other lots were kept over winter in a dry condition and planted in the spring in potting soil in a greenhouse. No seedlings have appeared, but a long rest period may be demanded and the seed may yet germinate.

EXPERIMENTS WITH DIVIDED RHIZOMES

In the spring of 1902, 40 plants were secured and planted under a shade of temporary character, but the season was too far advanced to permit of much growth during that year. In 1903, proper shade was supplied, all other conditions were better, and the plants made a good growth. The crop was dug about the middle of November 1903; the roots were weighed and divided. They were again planted and in May, 1904, there were found to be 150 strong plants and a few smaller ones as a result of this division, an increase of 275 per cent.

This method of propagation seems to be the most important and the other two of second importance. The processes are simple and no skill is needed. The plant dies down in late summer and the stem decays, leaving a scar in its place on the rhizome. Two or more buds are formed on the sides of the rhizome and these accumulate energy for growth the following spring. If the root is cut in as many pieces as there are buds, giving each plant a portion of the rhizome, some fibrous roots, and one or more buds, the number of the plants can be doubled. The roots are planted and mulched and the process is complete. The rains pack the soil around the roots and they are ready to grow when spring comes. The process may be repeated every year and the number of roots increased indefinitely.

The stronger fibrous roots of the larger plants dug in the autumn of 1903 were formed. from a few inches to a foot from the rhizome. Some were about half an inch long, but the majority of them were smaller. The larger ones need no special treatment and may be planted with the main crop. The smaller ones should be planted in boxes or beds of well prepared soil, at a distance of about 3 inches apart, mulched with a thin coating of leaf mold or similar material, and grown in shade until large enough to transplant to the shelter with the larger plants. They will probably require at least three years to reach their full development. If they could be left undisturbed in the beds where they are formed they would receive nourishment from the older rhizomes and perhaps grow faster, but it is probably best to divide the older roots every year where propagation alone is desired, planting the smaller roots and the plants made by division of the rhizomes.

The larger roots are marketed to more advantage than the smaller ones, so it is best to have the surplus consist of the larger roots. The frequent working of the soil allowed by this treatment will keep it in better condition than if left undisturbed for a longer period.

YIELD OF ROOTS

The yield from the small plant grown by the Department was 4 pounds of green roots to an eighth of a square rod of soil, or 5,120 pounds per acre. This, when dried, would give about 1,500 pounds of marketable roots. The conditions were not very good, the shade being too close to the plants and the plants being set too far apart. The yield will probably be larger with the shade now in use. The 150 roots obtained by dividing the above crop now occupy less than one-fourth of a square rod and are set in rows one foot apart and 6 inches apart in the rows.

TIME NECESSARY TO MATURE CROP

The number of years necessary to produce the largest crop has not been definitely determined, but the roots begin to decay after the fourth year and the central and largest part of the root decays at the oldest scar, leaving two or more plants in place of the old one. No advantage can be gained by growing the plants more than three years and probably very little by growing them more than two years. For propagation alone, one year will give good results, while for maintaining a constant area and producing a crop, two or three years, depending upon the growth made, will give a good crop of large, marketable roots.

MARKET CONDITIONS

Golden Seal is a root the price of which has fluctuated widely, because of the alternate oversupply and scarcity, manipulation of the market, lack of demand, or other influences. High prices will cause the diggers to gather the root in abundance, thus overstocking the market, which the next season results in lower prices, at which diggers refuse to collect the root, thus again causing a shortage in the supply. Lack of demand usually brings about a shrinkage in price, even tho the supply is light, while an active demand will cause prices to advance in spite of a plentiful supply.

The arrival of spring dug root has a weakening effect on the market, altho the fall dug root is always preferred. For the past few years, however, high prices have been steadily maintained and there appears to be but one cause for this and that is, as already pointed out, that the forests no longer yield unlimited quantities of this valuable root, as in former years, and the scant supply that can be had is inadequate to meet the constantly increasing demand. According to the market reports contained in the Oil, Paint and Drug Reporter, the year 1904 opened with a quotation of 74 to 75 cents, will soon advance (in one week early in February) from 76 cents to 95 cents. A still further advance occurred about the end of February, when the price went up from $1.00 to $1.25 per pound.

In March the market was almost destitute of supplies, but lack of interest brought the price down to $1.10. In May the price again advanced to $1.25 and it was stated that the local supplies were being held by a small number of dealers, altho it was believed that together they held not more than 1,000 pounds. About June 1st the arrival of spring dug roots cause a the market to sag, prices ranging from $1.10 to $1.18 during that month and in July from 90 cents to $1.10. In August the lowest price was $1.15 and the highest $1.50, no discrimination being made between the fall dug and the spring dug roots. From September 1st to October 15th, 1904, the price of Golden Seal varied but little, $1.35 being the lowest and $1.40 the highest quotation. No supplies worth mentioning can be obtained in the West; the stock in New York is short and the demand, especially for export, is increasing. It is impossible to ascertain the exact annual consumption of Golden Seal root, but the estimates furnished by reliable dealers place these figures at from 200,000 to 300,000 pounds annually, about one-tenth of which is probably used for export. It will be observed that the price of this article is very sensitive to market conditions and it seems probable that the point of overproduction would be easily reached if a large number of Golden Seal growers were to meet with success in growing large areas of this drug.

2

Tea: An Introduction

Tea is an aromatic beverage commonly prepared by pouring hot or boiling water over cured leaves of the Camellia sinensis, an evergreen shrub native to Asia. After water, it is the most widely consumed drink in the world. Some teas, like Darjeeling and Chinese greens, have a cooling, slightly bitter, and astringent flavour, while others have vastly different profiles that include sweet, nutty, floral, or grassy notes.

Tea originated in the southwest of China, possibly as a medicinal drink. Portuguese priests and merchants introduced it to the West during the 16th century. During the 17th century, drinking tea became fashionable among Britons, who started large-scale production and commercialization of the plant in India to bypass a Chinese monopoly at that time. The phrase herbal tea usually refers to infusions of fruit or herbs made without the tea plant, such as steeps of rosehip, chamomile, or rooibos. These are also known as tisanes or herbal infusions to distinguish them from "tea" as it is commonly construed.

The Chinese character for tea is originally written with an extra stroke as (pronounced tu, used as a word for a bitter herb), and acquired its current form during the Tang Dynasty as used in the eighth-century treatise on tea, The Classic of Tea. The word is pronounced differently in the different varieties of Chinese, such as chá in Mandarin, zo and dzo in Wu Chinese, and ta and te in Min Chinese. One suggestion is that the different pronunciations may have arisen from the different words for tea in ancient China, for example tu may have given rise to tê; historical phonologists however argued that the cha, te and dzo all arose from the same root with a reconstructed pronunciation dra (dr- represents a single consonant for a retroflex d), which changed due to sound shift through the centuries. Other ancient words for tea include jia (defined as "bitter tu" during the Han Dynasty), she, ming and chuan, with ming the only other word still in common use for tea. It has been proposed that the Chinese words for tea, tu, cha and ming, are themselves also words borrowed from the Austro-Asiatic languages of people who originally inhabited the botanical homeland of tea plant in southwest China; cha for example may have been derived from an archaic Austro-Asiatic root *la, meaning "leaf". Most Chinese languages, such as Mandarin and Cantonese, pronounce it along the lines of cha, but Hokkien varieties along the Southern coast of China and in Southeast Asia pronounce it like teh. These two pronunciations have made their separate ways into other languages around the world:

- Te is from the Amoy tê of southern Fujian province. It reached the West from the port of Xiamen (Amoy), once a major point of contact with Western European traders such as the Dutch, who spread it to Western Europe. This pronunciation gives rise to English "tea" and other similar words in other languages, and is the most common form worldwide.
- Cha is from the Cantonese chàh of Guangzhou (Canton) and the ports of Hong Kong and Macau, also major points of contact, especially with the Portuguese, who spread it to India in the 16th century. The Korean and Japanese pronunciations of cha, however, came not from Cantonese, rather they were borrowed into Korean and Japanese during earlier periods of Chinese history.

A third form, the increasingly widespread chai came from Persian chay. Both the châ and chây forms are found in Persian dictionaries. They derive from Northern Chinese pronunciation of chá, which passed overland to Central Asia and Persia, where it picked up the Persian grammatical suffix -yi before passing on to Russian, Arabic, Urdu, Turkish, etc.

English has all three forms: cha or char (both pronounced), attested from the 16th century; tea, from the 17th; and chai, from the 20th.

Languages in more intense contact with Chinese, Sinospheric languages such as Vietnamese, Zhuang, Tibetan, Korean, and Japanese, may have borrowed their words for tea at an earlier time and from a different variety of Chinese, so-called Sino-Xenic pronunciations. Although normally pronounced as cha, Japanese also retains the early but now uncommon pronunciations of ta and da, similarly Korean also has ta in addition to cha, and Vietnamese trà in addition to chà. Japanese has different pronunciations for the word tea depending on when the pronunciations were first borrowed into the language: Ta comes from the Tang Dynasty court at Chang'an—that is, from Middle Chinese; da, however, comes from the earlier Southern Dynasties court at Nanjing, a place where the consonant was still voiced, as it is today in neighbouring Shanghainese zo.

ORIGIN AND HISTORY

The history of tea is long and complex, spreading across multiple cultures over the span of thousands of years. Tea likely originated in southwest China during the Shang dynasty as a medicinal drink. An early credible record of tea drinking dates to the 3rd century AD, in a medical text written by Hua Tuo. Tea was first introduced to Portuguese priests and merchants in China during the 16th century. Drinking tea became popular in Britain during the 17th century. The British introduced tea production, as well as tea consumption, to India, in order to compete with the Chinese monopoly on tea.

GEOGRAPHIC ORIGINS

- "Camellia sinensis originated in southeast Asia, specifically around the intersection of latitude 29°N and longitude 98°E, the point of confluence of the lands of northeast India, north Burma, southwest China and Tibet. The plant was introduced to more than 52 countries, from this 'centre of origin'."

Based on morphological differences between the Assamese and Chinese varieties, botanists have long asserted a dual botanical origin for tea; however, statistical cluster analysis, the same chromosome number (2n=30), easy hybridization, and various types of intermediate hybrids and spontaneous polyploids all appear to demonstrate a single place of origin for Camellia sinensis — the area including the northern part of Burma, and Yunnan and Sichuan provinces of China.

Yunnan Province has also been identified as "the birthplace of tea...the first area where humans figured out that eating tea leaves or brewing a cup could be pleasant." Fengqing County in the Lincang City Prefecture of Yunnan Province in China is said to be home to the world's oldest cultivated tea tree, some 3,200 years old.

According to The Story of Tea, tea drinking likely began in modern-day Yunnan province during the Shang Dynasty (1500 BC–1046 BC), as a medicinal drink. From there, the drink spread to Sichuan, and it is believed that there "for the first time, people began to boil tea leaves for consumption into a concentrated liquid without the addition of other leaves or herbs, thereby using tea as a bitter yet stimulating drink, rather than as a medicinal concoction."

ORIGIN MYTHS

In one popular Chinese legend, Shennong, the legendary Emperor of China and inventor of agriculture and Chinese medicine was drinking a bowl of just boiled water due to a decree that his subjects must boil water before drinking it some time around 2737 BC when a few leaves were blown from a nearby tree into his water, changing the color. The emperor took a sip of the brew and was pleasantly surprised by its flavor and restorative properties. A variant of the legend tells that the emperor tested the medical properties of various herbs on himself, some of them poisonous, and found tea to work as an antidote. Shennong is also mentioned in Lu Yu's famous early work on the subject, The Classic of Tea. A similar Chinese legend goes that the god of agriculture would chew the leaves, stems, and roots of various plants to discover medicinal herbs. If he consumed a poisonous plant, he would chew tea leaves to counteract the poison. A rather gruesome legend dates back to the Tang Dynasty. In the legend, Bodhidharma, the founder of Chan Buddhism, accidentally fell asleep after meditating in front of a wall for nine years. He woke up in such disgust at his weakness that he cut off his own eyelids. They fell to the ground and took root, growing into tea bushes. Sometimes, another version of the story is told with Gautama Buddha in place of Bodhidharma.

Scholars however believe that tea drinking likely originated in the southwest of China, and that the Chinese words for tea themselves may have been originally derived from the Austro-Asiatic languages of the people who originally inhabited that area. Whether or not these legends have any basis in fact, tea has played a significant role in Asian culture for centuries as a staple beverage, a curative, and a status symbol. It is not surprising, therefore, that theories of its origin are often religious or royal in nature.

EARLY HISTORY

China

The Chinese have consumed tea for thousands of years. People of the Han Dynasty used tea as medicine (though the first use of tea as a stimulant is unknown). China is considered to have the earliest records of tea consumption, with possible records dating back to the 10th century BC. Note however that the current word for tea in Chinese only came into use in the 8th century AD, there are therefore uncertainties as to whether the older words used are the same as tea. The word tu appears in Shijing and other ancient texts to signify a kind of "bitter vegetable", and it is possible that it referred to a number of different plant, such as sowthistle, chicory, or smartweed, perhaps including tea. In the Chronicles of Huayang, it was recorded that the Ba people in Sichuan presented tu to the Zhou king. The state of Ba and its neighbour Shu were later conquered by the Qin, and according to the 17th century scholar Gu Yanwu who wrote in Ri Zhi Lu: "It was after the Qin had taken Shu that they learned how to drink tea."

The first known reference to boiling tea came from the Han dynasty work "The Contract for a Youth" written by Wang Bao where, among the tasks listed to be undertaken by the youth, "he shall boil tea and fill the utensils and "he shall buy tea at Wuyang". The first record of cultivation of tea also dated it to this period (Ganlu era of Emperor Xuan of Han) when tea was cultivated on Meng Mountain near Chengdu. From the Tang to the Qing dynasties, the first 360 leaves of tea grown here were picked each spring and presented to the emperor. Even today its green and yellow teas, such as the Mengding Ganlu tea, are still sought after. An unknown Chinese inventor was also the first person to invent a tea shredder. An early credible record of tea drinking dates to 220 AD, in a medical text Shi Lun by Hua Tuo, who stated, "to drink bitter t'u constantly makes one think better." Another possible early reference to tea is found in a letter written by the Qin Dynasty general Liu Kun. However, before the mid-8th century Tang dynasty, tea-drinking was primarily a southern Chinese practice. It became widely popular during the Tang Dynasty, when it was spread to Korea, Japan, and Vietnam.

Laozi, the classical Chinese philosopher, described tea as "the froth of the liquid jade" and named it an indispensable ingredient to the elixir of life. Legend has it that master Lao was saddened by society's moral decay and, sensing that the end of the dynasty was near, he journeyed westward to the unsettled territories, never to be seen again. While passing along the nation's border, he encountered and was offered tea by a customs inspector named Yin Hsi. Yin Hsi encouraged him to compile his teachings into a single book so that future generations might benefit from his wisdom. This then became known as the Dao De Jing, a collection of Laozi's sayings.

During the Sui Dynasty (589–618 AD). tea was introduced to Japan by Buddhist monks.

The Tang Dynasty writer Lu Yu's Cha Jing (The Classic of Tea) is an early work on the subject. According to Cha Jing tea drinking was widespread. The book describes how tea plants were grown, the leaves processed, and tea prepared as a beverage. It also describes how tea was evaluated. The book also discusses where the best tea leaves were produced. Teas produced in this period were mainly tea bricks which were often used as currency, especially further from the center of the empire where coins lost their value.

During the Song Dynasty (960–1279), production and preparation of all tea changed. The tea of Song included many loose-leaf styles (to preserve the delicate character favored by court society), but a new powdered form of tea emerged. Steaming tea leaves was the primary process used for centuries in the preparation of tea. After the transition from compressed tea to the powdered form, the production of tea for trade and distribution changed once again. The Chinese learned to process tea in a different way in the mid-13th century. Tea leaves were roasted and then crumbled rather than steamed. This is the origin of today's loose teas and the practice of brewed tea.

During the Yuan and Ming dynasties, unfermented tea leaves were first pan-fried, then rolled and dried. This stops the oxidation process which turns the leaves dark and allows tea to remain green. In the 15th century, Oolong tea, where the tea leaves were allowed to partially ferment before pan-frying, was developed.

Western taste, however, preferred the fully oxidized black tea, and the leaves were allowed to ferment further. Yellow tea was an accidental discovery in the production of green tea during the Ming dynasty, when apparently sloppy practices allowed the leaves to turn yellow, but yielded a different flavour as a result.

Tea production in China, historically, was a laborious process, conducted in distant and often poorly accessible regions. This led to the rise of many apocryphal stories and legends surrounding the harvesting process. For example, one story that has been told for many years is that of a village where monkeys pick tea. According to this legend, the villagers stand below the monkeys and taunt them. The monkeys, in turn, become angry, and grab handfuls of tea leaves and throw them at the villagers. There are products sold today that claim to be harvested in this manner, but no reliable commentators have observed this firsthand, and most doubt that it happened at all. For many hundreds of years the commercially used tea tree has been, in shape, more of a bush than a tree. "Monkey picked tea" is more likely a name of certain varieties than a description of how it was obtained. In 1391, the Ming court issued a decree that only loose tea would be accepted as a "tribute". As a result, loose tea production increased and processing techniques advanced. Soon, most tea was distributed in full-leaf, loose form and steeped in earthenware vessels.

Hong Kong

In Hong Kong, apart from the yum cha culture of southern China, a localised version of English tea was developed, the Hong Kong-style milk tea.

Japan

Tea use spread to Japan about the sixth century. Tea became a drink of the religious classes in Japan when Japanese priests and envoys, sent to China to learn about its culture, brought tea to Japan. Ancient recordings indicate the first batch of tea seeds were brought by a priest named Saichô (767–822) in 805 and then by another named Kûkai (774–835) in 806. It became a drink of the royal classes when Emperor Saga, the Japanese emperor, encouraged the growth of tea plants. Seeds were imported from China, and cultivation in Japan began.

In 1191, the famous Zen priest Eisai (1141–1215) brought back tea seeds to Kyoto. Some of the tea seeds were given to the priest Myoe Shonin, and became the basis for Uji tea. The oldest tea specialty book in Japan, Kissa Yôjôki (How to Stay Healthy by Drinking Tea), was written by Eisai. The two-volume book was written in 1211 after his second and last visit to China. The first sentence states, "Tea is the ultimate mental and medical remedy and has the ability to make one's life more full and complete." Eisai was also instrumental in introducing tea consumption to the warrior class, which rose to political prominence after the Heian Period.

Green tea became a staple among cultured people in Japan—a brew for the gentry and the Buddhist priesthood alike. Production grew and tea became increasingly accessible, though still a privilege enjoyed mostly by the upper classes. The tea ceremony of Japan was introduced from China in the 15th century by Buddhists as a semi-religious social custom. The modern tea ceremony developed over several centuries by Zen Buddhist monks under the original guidance of the monk Sen no Rikyû (1522–1591). In fact, both the beverage and the ceremony surrounding it played a prominent role in feudal diplomacy. In 1738, Soen Nagatani developed Japanese sencha, literally roasted tea, which is an unfermented form of green tea. It is the most popular form of tea in Japan today. In 1835, Kahei Yamamoto developed gyokuro, literally jewel dew, by shading tea trees during the weeks leading up to harvesting. At the end of the Meiji period (1868–1912), machine manufacturing of green tea was introduced and began replacing handmade tea.

Korea

The first historical record documenting the offering of tea to an ancestral god describes a rite in the year 661 in which a tea offering was made to the spirit of King Suro, the founder of the Geumgwan Gaya Kingdom (42–562). Records from the Goryeo Dynasty (918–1392) show that tea offerings were made in Buddhist temples to the spirits of revered monks.

During the Joseon Dynasty (1392–1910), the royal Yi family and the aristocracy used tea for simple rites. The "Day Tea Rite" was a common daytime ceremony, whereas the "Special Tea Rite" was reserved for specific occasions. Toward the end of the Joseon Dynasty, commoners joined the trend and used tea for ancestral rites, following the Chinese example based on Zhu Xi's text formalities of Family.Stoneware was common, ceramic more frequent, mostly made in provincial kilns, with porcelain rare, imperial porcelain with dragons the rarest. The earliest kinds of tea used in tea ceremonies were heavily pressed cakes of black tea, the equivalent of aged pu-erh tea still popular in China. However, importation of tea plants by Buddhist monks brought a more delicate series of teas into Korea, and the tea ceremony. Green tea, "Jakseol" or "Jungno", is most often served. However, other teas such as "Byeoksoryeong" Cheonhachun, Ujeon, Okcheon, as well as native chrysanthemum tea, persimmon leaf tea, or mugwort tea may be served at different times of the year.

Vietnam

Vietnamese green teas have been largely unknown outside of mainland Asia until the present day. Recent free-enterprise initiatives are introducing these green teas to outside countries through new export activities. Some specialty Vietnamese teas include Lotus tea and Jasmine tea. Vietnam also produces black and oolong teas in lesser quantities. Vietnamese teas are produced in many areas that have been known for tea-house "retreats." For example, some are located amidst immense tea forests of the Lamdong highlands, where there is a community of ancient Ruong houses built at the end of the 18th century.

GLOBAL EXPANSION

The earliest record of tea in a more occidental writing is said to be found in the statement of an Arabian traveler, that after the year 879 the main sources of revenue in Canton were the duties on salt and tea. Marco Polo records the deposition of a Chinese minister of finance in 1285 for his arbitrary augmentation of the tea taxes. The travelers Giovanni Batista Ramusio (1559), L. Almeida (1576), Maffei (1588), and Teixeira (1610) also mentioned tea. In 1557, Portugal established a trading port in Macau and word of the Chinese drink "chá" spread quickly, but there is no mention of them bringing any samples home.

In the early 17th century, a ship of the Dutch East India Company brought the first green tea leaves to Amsterdam from China. Tea was known in France by 1636. It enjoyed a brief period of popularity in Paris around 1648. The history of tea in Russia can also be traced back to the seventeenth century. Tea was first offered by China as a gift to Czar Michael I in 1618. The Russian ambassador tried the drink; he did not care for it and rejected the offer, delaying tea's Russian introduction by fifty years. In 1689, tea was regularly imported from China to Russia via a caravan of hundreds of camels traveling the year-long journey, making it a precious commodity at the time. Tea was appearing in German apothecaries by 1657 but never gained much esteem except in coastal areas such as Ostfriesland. Tea first appeared publicly in England during the 1650s, where it was introduced through coffeehouses. From there it was introduced to British colonies in America and elsewhere.

Portugal

Tea was first introduced to Portuguese priests and merchants in China during the 16th century, at which time it was termed chá.

Greece and Cyprus

Throughout Greece and Cyprus Greek tea is made with cinnamon and cloves.

India

Tea was first introduced into India by the British, in an attempt to break the Chinese monopoly on tea. The British, "using Chinese seeds, plus Chinese planting and cultivating techniques, launched a tea industry by offering land in Assam to any European who agreed to cultivate tea for export." Tea was originally only consumed by Anglicized Indians, it was not until the 1950s that tea grew widely popular in India through a successful advertising campaign by the India Tea Board.Prior to the British, the plant may have been used for medicinal purposes. Some cite the Sanjeevani tea plant first recorded reference of tea use in India. However, scientific studies have shown that the Sanjeevani plant is in fact a different plant and is not related to tea. The Singpho tribe and the Khamti tribe also validate that they have been consuming tea since the 12th century. However, commercial production of tea in India did not begin until the arrival of the British East India Company, at which point large tracts of land were converted for mass tea production.

The Chinese variety is used for Sikkim, Darjeeling tea, and the Assamese variety, clonal to the native to Assam, everywhere else. The British started commercial tea plantations in India and in Ceylon: "In 1824 tea plants were discovered in the hills along the frontier between Burma and Assam. The British introduced tea culture into India in 1836 and into Ceylon (Sri Lanka) in 1867. At first they used seeds from China, but later seeds from the clonal Assam plant were used." Only black tea was produced until recent decades. India was the top producer of tea for nearly a century, but was displaced by China as the top tea producer in the 21st century. Indian tea companies have acquired a number of iconic foreign tea enterprises including British brands Tetley and Typhoo. While India is the largest consumer of tea worldwide, the per-capita consumption of tea in India remains a modest 750 grams per person every year. Recently consumption of green tea has seen a great upsurge across the cities. Average growth in the consumption is assumed to be over 50%. One estimate suggest the market size has already crossed over INR 1400crore and will reach a 6000 crore in next few years. Top station, 41 km (1 Hour) from Munnar, is aptly named, as it is home to some of the highest tea plantations in India. It lies on the state of Kerala and commands a panoramic view of rolling green hills.

Iran

Gilan in North of Iran is main production center of Iranian Tea. Historically, Lahijan is the first town in Iran to have tea plantations. With its mild weather, soil quality and fresh spring water, Lahijan stands to have the largest area of tea cultivation in Iran. "Lahijan Spring Tea" is the best quality tea produced in the country. Tea is cultivated at other cities of Gilan, for example Fuman and Roudsar.

Taiwan

Taiwan is famous for the making of Oolong tea and green tea, as well as many western-styled teas. Bubble Tea or "Zhen Zhu Nai Cha" is black tea mixed with sweetened condensed milk and tapioca. Since the island was known to Westerners for many centuries as Formosa — short for the Portuguese Ilha Formosa, or "beautiful island" — tea grown in Taiwan is often identified by that name.

Thailand

Thai tea or "cha-yen" in Thailand, is a drink made from strongly brewed black tea ("red tea" in East Asia). Other ingredients may include added orange blossom water, star anise, crushed tamarind seed or red and yellow food coloring, and sometimes other spices as well. This tea is sweetened with sugar and condensed milk.

Usually, Thai people drink Thai hot tea in the morning, frequently with Yau ja gwai (fried dough) or Pa-tong-ko.

The varieties of Thai tea include:

- Thai hot tea is Thai tea served hot.
- Dark Thai hot tea is Thai tea served hot with no milk content, sweetened with sugar only.
- Dark Thai iced tea is Thai tea served cold with ice and without milk.

Turkey

Turkey is traditionally one of the largest tea markets in the world. Turkish black tea is the most popular drink in Turkey, even more popular than Turkish coffee.

United Kingdom

The first record of tea in English came from a letter written by Richard Wickham, who ran an East India Company office in Japan, writing to a merchant in Macao requesting "the best sort of chaw" in 1615. Peter Mundy, a traveller and merchant who came across tea in Fujian in 1637, wrote, "chaa — only water with a kind of herb boyled in it ". In 1657, tea was listed as an item in the price list in a London coffee house, and the first advertisement for tea appear in 1658. On 25 September 1660 Samuel Pepys recorded in his diary: "I did send for a cup of tee (a China drink) of which I never had drank before." It is probable that early imports were smuggled via Amsterdam or through sailors arriving on eastern boats. The marriage of King Charles II in 1662 to the Portuguese princess Catherine of Braganza also brought the tea drinking habit to court. Official trade of tea began in the 1660s at with an import of only 2 ounces, but grew to 24 million pounds a year by 1801.

Regular trade began in Canton (now Guangzhou), where it was controlled by two monopolies: the Chinese Cohong (trading companies) and the British East India Company. The Cohong acquired tea from 'tea men' who had an elaborate supply chain into the mountains and provinces where tea grew.

The East India Company brought back many products, of which tea was just one, but it was to prove one of the most successful. It was initially promoted as a medicinal beverage or tonic but by the end of the seventeenth century was taken as an all-purpose drink, albeit mainly by the elite as it was an expensive drink. Tea was not traded in significant amount until the 18th century. Black tea overtook green tea in popularity in the 1720s and British drinkers added sugar and milk to tea, a practice that was not done in China. Tea became increasingly popular and by 1750 had become the national drink. The origin of large trade in tea was the need for a return cargo from the East Indies. Merchant ships delivered fabrics manufactured in Britain to India and China but would return empty or partially full. To solve this problem the East India Company began a vigorous public relations campaign in England to popularize tea among the common people in Britain and develop it as a viable return cargo. A fungus reduced Ceylon's coffee production by 95% in the 19th century, cementing tea's popularity.

The escalation of tea importation and sales over the period 1690 to 1750 is mirrored closely by the increase in importation and sales of cane sugar: the British were not drinking just tea but sweet tea. Thus, two of Britain's trading triangles converged: the sugar sourced from Britain's trading triangle encompassing Britain, Africa and the West Indies and the tea from the triangle encompassing Britain, India and China.

In China, the Qing dynasty Qianlong Emperor decreed that "China was the center of the world and had everything they could ever need, so all trade with foreigners must be paid for in silver!" This meant that British traders had to pay China for its tea with silver bullion. Critics of the tea trade at this time would point to the damage caused to Britain's wealth by this loss of bullion. As a way to generate the silver needed as payment for tea, Britain began exporting opium from the traditional growing regions of British India (in present-day Pakistan and Afghanistan) into China. Opium use in China had a long history however British importation of opium, which began in the late 18th century, increased fivefold between 1821 and 1837, and the Qing government attitude towards opium, which was often ambivalent, hardened as usage of the drug spread more widely across Chinese society. It finally began serious measures to curtail imports in 1838–39. Tea by now had become an important source of tax revenue for the British Empire and the banning of the opium trade and thus the creation of funding issues for tea importers was one of the main causes of the First Opium War.

Whilst waging war on China was one of Britain's tactics it also began to explore, then executed, a plan to use India for growing tea. After tea plants were smuggled out of China, plantations were established in areas such as Darjeeling, Assam, and Ceylon. As an attempt to circumvent its dependence on Chinese tea, the East India Company sent Scottish botanist Robert Fortune to China to purchase and bring out of China tea plants, which were then taken to India, although it was the discovery of native varieties of tea plant in India which proved more important for the development of production there. Tea remained a very important item in Britain's global trade, contributing in part to Britain's global dominance by the end of the eighteenth century. To this day tea is seen worldwide as a symbol of 'Britishness', but also, to some, as a symbol of old British colonialism. The London 2012 section of the paralympic handover in Beijing included tea as part of the routine. A cup or mug of tea in Britain is usually made in a different way than is common in China and other Eastern countries. Over 90% of tea consumed is black tea, often but not always with a small amount of milk and / or sugar added. The tea used is often contained in a tea bag. As of 2009 the UK can boast one commercial tea plantation with another planned. The existing one lies in Cornwall and is owned by the Tregothnan Estate. By 2015 another will lie in Pembrokeshire, Wales, owned by the Pembrokeshire Tea Company.

United States

While coffee is by far more popular, hot brewed black tea is enjoyed both with meals and as a refreshment by much of the population. Similarly, iced tea is consumed throughout. In the Southern states sweet tea, sweetened with large amounts of sugar or an artificial sweetener and chilled, is the fashion. Outside the South, sweet tea is sometimes found, but primarily because of cultural migration and commercialization. The drinking of tea in the United States was largely influenced by the passage of the Tea Act and its subsequent protest during the American Revolution. Tea consumption sharply decreased in America during and after the Revolution, when many Americans switched from drinking tea to drinking coffee, considering tea drinking to be unpatriotic. The American specialty tea market quadrupled in the years from 1993 to 2008, now being worth $6.8 billion a year. Similar to the trend of better coffee and better wines, this tremendous increase was partly due to consumers who choose to trade up. Specialty tea houses and retailers also started to pop up during this period.

Canada

Until the Second World War, Canadians were big tea drinkers since the days of British colonization. As a result of the war, Canadians drank more coffee like their American neighbors to the south. During the 1990s, Canadians begun to purchase more specialty teas instead of coffee. A commercial tea farm has opened on Vancouver Island in British Columbia in 2010 and expects production to start in 2015.

Australia

The Aboriginal Australians drank an infusion from the plant species leptospermum (a different plant from the tea plant or camellia sinensis). Upon discovering Australia, Captain Cook noticed the aboriginal peoples drinking it and called it tea. Today the plant is referred to as the "ti tree." Through colonisation by the British, tea was introduced to Australia. In fact, tea was aboard the First Fleet in 1788. Tea is a large part of modern Australian culture due to its British origins. Australians drink tea and have afternoon tea and morning tea much the way the British do. Additionally, due to Australia's climate, tea is able to be grown and produced in northern Australia. In 2000, Australia consumed 14,000 tonnes of tea annually. Tea production in Australia remains very small and is primarily in northern New South Wales and Queensland. Most tea produced in Australia is black tea, although there are small quantities of green tea produced in the Alpine Valleys region of Victoria.

In 1884, the Cutten brothers established the first commercial tea plantation in Australia in Bingil Bay in northern Queensland. In 1883, Alfred Bushell opened the first tea shop in Australia in present-day Queensland. In 1899, Bushell's sons moved the enterprise to Sydney and began selling tea commercially, founding Australia's first commercial tea seller Bushell's Company.

Sri Lanka

Sri Lanka is renowned for its high quality tea and as the fourth biggest tea producing country globally, after China, India and Kenya, and has a production share of 9% in the international sphere. The total extent of land under tea cultivation has been assessed at approximately 187,309 hectares. The plantations started by the British were initially taken over by the government in the 1960s, but have been privatized and are now run by 'plantation companies' which own a few 'estates' or tea plantations each. Ceylon tea is divided into 3 groups as Upcountry, Mid country and Low country tea based on the geography of the land on which it is grown.

Africa and South America

Africa and South America have seen greatly increased tea production in recent decades, the great majority for export to Europe and North America respectively, produced on large estates, often owned by tea companies from the export markets. Almost all production is of basic mass-market teas, processed by the Crush, Tear, Curl method. Kenya is now the third largest global producer, after China and India, and is now the largest exporter of tea to the United Kingdom. There is also a great consumption of tea in Chile. In South Africa, the non-Camellia sinensis beverage rooibos is popular. In South America yerba mate is a popular infused beverage. The only European plantation is Chá Gorreana, located in Ribeira Grande, São Miguel island, Azores (Portugal).

CULTIVATION AND HARVESTING

Camellia sinensis is an evergreen plant that grows mainly in tropical and subtropical climates. Some varieties can also tolerate marine climates and are cultivated as far north as Cornwall in the United Kingdom, Perthshire in Scotland, Washington state in the United States, Vancouver Island in Canada,. In the Southern Hemisphere, tea is grown as far south as Hobart on the Australian island of Tasmania and Waikato in New Zealand.

Tea plants are propagated from seed and cuttings; about 4 to 12 years are needed for a plant to bear seed and about three years before a new plant is ready for harvesting. In addition to a zone 8 climate or warmer, tea plants require at least 127 cm (50 in) of rainfall a year and prefer acidic soils. Many high-quality tea plants are cultivated at elevations of up to 1,500 m (4,900 ft) above sea level. While at these heights the plants grow more slowly, they acquire a better flavour.

Two principal varieties are used: Camellia sinensis var. sinensis, which is used for most Chinese, Formosan and Japanese teas, and C. s. var. assamica, used in Pu-erh and most Indian teas (but not Darjeeling). Within these botanical varieties, many strains and modern clonal varieties are known. Leaf size is the chief criterion for the classification of tea plants, with three primary classifications being, Assam type, characterised by the largest leaves; China type, characterised by the smallest leaves; and Cambodian type, characterised by leaves of intermediate size.

A tea plant will grow into a tree of up to 16 m (52 ft) if left undisturbed, but cultivated plants are generally pruned to waist height for ease of plucking. Also, the short plants bear more new shoots which provide new and tender leaves and increase the quality of the tea.

Only the top 1–2 in of the mature plant are picked. These buds and leaves are called 'flushes'. A plant will grow a new flush every seven to 15 days during the growing season. Leaves that are slow in development tend to produce better-flavoured teas. Pests of tea include mosquito bugs of the genus Helopeltis (which are true bugs that must not be confused with the dipteran) that can tatter leaves, so they may be sprayed with insecticides.

PROCESSING AND CLASSIFICATION

Tea is generally divided into categories based on how it is processed. At least six different types are produced:

- *White:* Wilted and unoxidized
- *Yellow:* Unwilted and unoxidized, but allowed to yellow
- *Green:* Unwilted and unoxidized
- *Oolong:* Wilted, bruised, and partially oxidized
- *Black:* Wilted, sometimes crushed, and fully oxidized (called 'red tea' in China)
- *Post-Fermented:* Green tea that has been allowed to ferment/compost ('black tea' for the Chinese)

The most common are white, green, oolong, and black.

After picking, the leaves of C. sinensis soon begin to wilt and oxidize unless immediately dried. An enzymatic oxidation process triggered by the plant's intracellular enzymes causes the leaves to turn progressively darker as their chlorophyll breaks down and tannins are released. This darkening is stopped at a predetermined stage by heating, which deactivates the enzymes responsible. In the production of black teas, halting by heating is carried out simultaneously with drying. Without careful moisture and temperature control during manufacture and packaging, growth of undesired molds and bacteria may make tea unfit for consumption.

BLENDING AND ADDITIVES

Although single-estate teas are available, almost all tea in bags and most loose tea sold in the West is blended. Such teas may combine others from the same cultivation area or several different ones. The aim is to obtain consistency, better taste, higher price, or some combination of the three.

Tea easily retains odors, which can cause problems in processing, transportation, and storage. This same sensitivity also allows for special processing (such as tea infused with smoke during drying) and a wide range of scented and flavoured variants, such as bergamot (found in Earl Grey), vanilla, and spearmint.

NUTRIENTS AND PHYTOCHEMICALS

Black and green teas contain no essential nutrients in significant content, with the exception of the dietary mineral, manganese at 0.5 mg per cup or 26% of the Daily Value. Tea leaves contain diverse polyphenols, including flavonoids, epigallocatechin gallate (commonly noted as EGCG) and other catechins. It has been suggested that green and black tea may protect against cancer or other diseases such as obesity or Alzheimer's disease, but the compounds found in green tea have not been conclusively demonstrated to have any effect on human diseases. One human study demonstrated that regular consumption of black tea over four weeks had no beneficial effect in lowering blood cholesterol levels.

TEA CULTURE

Tea may be consumed early in the day to heighten calm alertness; it contains L-theanine, theophylline, and bound caffeine (sometimes called theine). Decaffeinated brands are also sold. While herbal teas are also referred to as tea, most of them do not contain leaves from the tea plant.

While tea is the second most consumed beverage on Earth after water, in many cultures it is also consumed at elevated social events, such as afternoon tea and the tea party. Tea ceremonies have arisen in different cultures, such as the Chinese and Japanese tea ceremonies, each of which employs traditional techniques and ritualised protocol of brewing and serving tea for enjoyment in a refined setting. One form of Chinese tea ceremony is the Gongfu tea ceremony, which typically uses small Yixing clay teapots and oolong tea. Turkish tea is an important part of Turkish cuisine, and is the most commonly consumed hot drink, despite the country's long history of coffee consumption. In 2004 Turkey produced 205,500 tonnes of tea (6.4% of the world's total tea production), which made it one of the largest tea markets in the world, with 120,000 tons being consumed in Turkey, and the rest being exported. In 2010 Turkey had the highest per capita consumption in the world at 2.7 kg. As of 2013, the per-capita consumption of Turkish tea exceeds 10 cups per day and 13.8 kg per year. Tea is grown mostly in Rize Province on the Black Sea coast.

Ireland has, for a long time, been one of the biggest per-capita consumers of tea in the world. The national average is four cups per person per day, with many people drinking six cups or more. Tea in Ireland is usually taken with milk or sugar and is slightly spicier and stronger than the traditional English blend. The two main brands of tea sold in Ireland are Lyons and Barry's. The Irish love of tea is perhaps best illustrated by the stereotypical housekeeper, Mrs. Doyle in the popular sitcom Father Ted.

Tea is prevalent in most cultures in the Middle East. In Arab culture, tea is a focal point for social gatherings. In Pakistan, tea is called chai. Both black and green teas are popular and are known locally as sabz chai and kahwah, respectively. The popular green tea called kahwah is often served after every meal in the Pashtun belt of Balochistan and in Khyber Pakhtunkhwa, which is where the Khyber Pass of the Silk Road is found. In the transnational Kashmir region, which straddles the border between India and Pakistan, Kashmiri chai or noon chai, a pink, creamy tea with pistachios, almonds, cardamom, and sometimes cinnamon, is consumed primarily at special occasions, weddings, and during the winter months when it is sold in many kiosks.

In central and southern Punjab and the metropolitan Sindh region of Pakistan, tea with milk and sugar (sometimes with pistachios, cardamom, etc.), commonly referred to as chai, is widely consumed. It is the most common beverage of households in the region. In the northern Pakistani regions of Chitral and Gilgit-Baltistan, a salty, buttered Tibetan-style tea is consumed. In Iranian culture, tea is so widely consumed, it is generally the first thing offered to a household guest.

In India, tea is one of the most popular hot beverages. It is consumed daily in almost all homes, offered to guests, consumed in high amounts in domestic and official surroundings, and is made with the addition of milk with or without spices. It is also served with biscuits dipped in the tea and eaten before consuming the tea. More often than not, it is drunk in "doses" of small cups (referred to as "Cutting" chai if sold at street tea vendors) rather than one large cup. On 21 April 2012, the Deputy Chairman of Planning Commission (India), Montek Singh Ahluwalia, said tea would be declared as national drink by April 2013. The move is expected to boost the tea industry in the country. Speaking on the occasion, Assam Chief Minister Tarun Gogoi said a special package for the tea industry would be announced in the future to ensure its development.

In the United States, 80% of tea is consumed as iced tea. Sweet tea is native to the southeastern US, and is iconic in its cuisine. Switzerland has its own unique blend of iced tea, made with the basic ingredients like black tea, sugar, lemon juice and mint, but a variety of Alp herbs are also added to the concoction. Apart from classic flavours like lemon and peach, exotic flavours like jasmine and lemongrass are also very popular.

In the United Kingdom, it is consumed daily and often by a majority of people across the country, and indeed is perceived as one of Britain's cultural beverages. In British homes, it is customary good manners for a host to offer tea to guests soon after their arrival. Tea is generally consumed at home; outside the home in cafés. Afternoon tea with cakes on fine porcelain is a cultural stereotype, sometimes available in quaint tea-houses. In southwest England, many cafes serve a 'cream tea', consisting of scones, clotted cream, and jam alongside a pot of tea. Throughout the UK, 'tea' may also refer to the evening meal. In Burma (Myanmar), tea is consumed not only as hot drinks, but also as sweet tea and green tea known locally as laphet-yay and laphet-yay-gyan, respectively. Pickled tea leaves, known locally as laphet, are also a national delicacy. Pickled tea is usually eaten with roasted sesame seeds, crispy fried beans, roasted peanuts and fried garlic chips.

PREPARATION

STEEPING TEA

The traditional method of preparing tea is to place loose tea leaves directly (or in a tea infuser) into a tea pot or teacup, pour freshly boiled water over the leaves, and allow the infused liquid to steep (or "brew"). After a few minutes, the infuser is removed, or the tea is poured through a strainer while serving. Strength should be varied by the amount of tea leaves used, not changing the steeping time. Most green teas should be allowed two or three minutes, although other types may vary between thirty seconds and ten minutes. Quantity also varies by tea type, with a basic recipe calling for one slightly heaped teaspoon (about 5 ml) for each teacup of water (200–240 ml) (7–8 oz). Stronger teas to be drunk with milk (such as Assam) are often prepared more heavily, while more delicate high-grown varieties (such as a Darjeeling) more lightly. Optimum brewing temperature depends on tea type. Camellia sinensis naturally contains tannins having bitter properties accentuated by both temperature and steeping time. These tannins are enhanced by oxidation during processing. Teas with little or no oxidation, such as a green or white, are best at lower temperatures between 65 and 85 °C (149 and 185 °F), while more oxidized teas require 100 °C (212 °F) to extract their large, complex, flavourful phenolic molecules. In addition, boiling reduces the dissolved oxygen content of water, which would otherwise react with phenolic molecules to degrade them.

Type	Water temp.	Steep time	Infusions
White tea	65 to 70 °C (149 to 158 °F)	1–2 minutes	3
Yellow tea	70 to 75 °C (158 to 167 °F)	1–2 minutes	3
Green tea	75 to 80 °C (167 to 176 °F)	1–2 minutes	4–6
Oolong tea	80 to 85 °C (176 to 185 °F)	2–3 minutes	4–6
Black tea	99 °C (210 °F)	2–3 minutes	2–3
Flowering tea	100 °C (212 °F)	2–3 minutes	4–5
Pu'er tea	95 to 100 °C (203 to 212 °F)	Limitless	Several
Tisanes	99 °C (210 °F)	3–6 minutes	Varied

Some tea sorts are often brewed several times using the same leaves. Historically in China, tea is divided into a number of infusions. The first is immediately poured out to wash the tea, and then the second and further infusions are drunk. The third through fifth are nearly always considered the best, although different teas open up differently and may require more infusions to produce the best flavour. One way to taste a tea throughout its entire process is to add hot water to a cup containing the leaves and sample it every 30 seconds. As the tea leaves unfold (known as "The Agony of the Leaves") the taste evolves. A tea cosy or a teapot warmer are often used to keep the temperature of the tea in a teapot constant over periods of 20–60 minutes.

COLD BREW TEA (AKA REFRIGERATOR TEA) AND SUN TEA

While most tea is prepared using hot water, it is also possible to brew a beverage from tea using room temperature or cooled water. This requires longer steeping time to extract the key components, and produces a different flavor profile. For best results, it is best to use about 1.5 times the tea leaves that would be used for hot steeping, and to refrigerate for 4-10 hours. The process of making cold brew tea is much simpler than that for cold brew coffee. Cold brewing has some disadvantages compared to hot steeping.

Firstly, if the leaves or source water contain unwanted bacteria, they may flourish, whereas using hot water has the benefit of killing most bacteria. This is less of a concern in modern times and developed regions. Secondly, cold brewing may allow for less caffeine to be extracted, which may or may not be desired. Sun Tea is made by steeping the tea leaves in a jar of unheated tap water left in the sun. It does not get hot enough to kill bacteria present on the tea leaves or in the water, such as Alcaligenes viscolactis.

BLACK TEA

Popular varieties of black tea include Assam, Iran, Nepal, Darjeeling, Nilgiri, Turkish, Keemun, and Ceylon teas. Many of the active substances in black tea do not develop at temperatures lower than 90 °C (194 °F). As a result, black tea in the West is usually steeped in water near its boiling point, at around 99 °C (210 °F). The most common fault when making black tea is to use water at too low a temperature. Since boiling point drops with increasing altitude, it is difficult to brew black tea properly in mountainous areas. Warming the tea pot before steeping is critical at any elevation.

Western black teas are usually brewed for about four minutes and are usually not allowed to steep for less than 30 seconds or more than about five minutes (a process known as brewing or mashing in Britain). In many regions of the world, however, actively boiling water is used and the tea is often stewed. In India, black tea is often boiled for fifteen minutes or longer to make Masala chai, as a strong brew is preferred. Tea should be strained while serving. A food safety management group of the International Organization for Standardization (ISO) has published a standard for preparing a cup of tea (ISO 3103: Tea — Preparation of liquor for use in sensory tests), primarily intended for standardizing preparation for comparison and rating purposes.

GREEN TEA

In regions of the world that prefer mild beverages, such as the West and Far East, green tea should be steeped in water around 80 to 85 °C (176 to 185 °F), the higher the quality of the leaves the lower the temperature. Regions such as North Africa or Central Asia prefer a bitter tea, and hotter water is used. In Morocco, green tea is steeped in boiling water for 15 minutes. The container in which green tea is steeped is often warmed beforehand to prevent premature cooling. High-quality green and white teas can have new water added as many as five or more times, depending on variety, at increasingly higher temperatures.

FLOWERING TEA

Flowering tea or blooming tea should be brewed at 100 °C (212 °F) in clear glass tea wares for up to three minutes. First pull 1/3 water to make the tea ball wet and after 30 seconds add the boiling water up to 4/5 of the capacity of the tea ware. The boiling water can help the tea ball bloom quickly and with a strong aroma of the tea. The height of glass tea ware should be 8–10 cm, which can help the tea and flowers bloom completely. One tea ball can be brewed 4-5 times.

OOLONG TEA

Oolong tea should be brewed around 185 to 205 °F, with the brewing vessel warmed before pouring the water. Yixing purple clay teapots are the traditional brewing-vessel for oolong tea which can be brewed multiple times from the same leaves, unlike green tea, seeming to improve with reuse. In the Chinese and Taiwanese Gongfu tea ceremony, the first brew is discarded, as it is considered a rinse of leaves rather than a proper brew.

PREMIUM OR DELICATE TEA

Some teas, especially green teas and delicate oolong teas, are steeped for shorter periods, sometimes less than 30 seconds. Using a tea strainer separates the leaves from the water at the end of the brewing time if a tea bag is not being used. However, the black Darjeeling tea, a premium Indian tea, needs a longer than average steeping time. Elevation and time of harvest offer varying taste profiles; proper storage and water quality also have a large impact on taste.

PU-ERH TEA

Pu-erh teas require boiling water for infusion. Some prefer to quickly rinse pu-erh for several seconds with boiling water to remove tea dust which accumulates from the ageing process, then infuse it at the boiling point (100 °C or 212 °F), and allow it to steep from 30 seconds to five minutes.

SERVING

To preserve the pretannin tea without requiring it all to be poured into cups, a second teapot may be used. The steeping pot is best unglazed earthenware; Yixing pots are the best known of these, famed for the high-quality clay from which they are made. The serving pot is generally porcelain, which retains the heat better. Larger teapots are a post-19th century invention, as tea before this time was very rare and very expensive. Experienced tea-drinkers often insist the tea should not be stirred around while it is steeping (sometimes called winding or mashing in the UK). This, they say, will do little to strengthen the tea, but is likely to bring the tannins out in the same way that brewing too long will do. For the same reason, one should not squeeze the last drops out of a teabag; if stronger tea is desired, more tea leaves should be used.

ADDITIVES

The addition of milk to tea in Europe was first mentioned in 1680 by the epistolist Madame de Sévigné. Many teas are traditionally drunk with milk in cultures where dairy products are consumed. These include Indian masala chai and British tea blends. These teas tend to be very hearty varieties of black tea which can be tasted through the milk, such as Assams, or the East Friesian blend. Milk is thought to neutralise remaining tannins and reduce acidity. The Han Chinese do not usually drink milk with tea but the Manchus do, and the elite of the Qing Dynasty of the Chinese Empire continued to do so. Hong Kong-style milk tea is based on British colonial habits. Tibetans and other Himalayan peoples traditionally drink tea with milk or yak butter and salt. In Eastern European countries (Russia, Poland and Hungary) and in Italy, tea is commonly served with lemon juice. In Poland, tea with milk is called a bawarka ("Bavarian style"), and is often drunk by pregnant and nursing women. In Australia, tea with milk is white tea.

The order of steps in preparing a cup of tea is a much-debated topic, and can vary widely between cultures or even individuals. Some say it is preferable to add the milk before the tea, as the high temperature of freshly brewed tea can denature the proteins found in fresh milk, similar to the change in taste of UHT milk, resulting in an inferior-tasting beverage. Others insist it is better to add the milk after brewing the tea, as black tea is often brewed as close to boiling as possible. The addition of milk chills the beverage during the crucial brewing phase, if brewing in a cup rather than using a pot, meaning the delicate flavour of a good tea cannot be fully appreciated. By adding the milk afterwards, it is easier to dissolve sugar in the tea and also to ensure the desired amount of milk is added, as the colour of the tea can be observed. Historically, the order of steps was taken as an indication of class: only those wealthy enough to afford good-quality porcelain would be confident of its being able to cope with being exposed to boiling water unadulterated with milk. Higher temperature difference means faster heat transfer so the earlier you add milk the slower the drink cools. A 2007 study published in the European Heart Journal found certain beneficial effects of tea may be lost through the addition of milk.

Many flavourings are added to varieties of tea during processing. Among the best known are Chinese jasmine tea, with jasmine oil or flowers, the spices in Indian masala chai, and Earl Grey tea, which contains oil of bergamot. A great range of modern flavours have been added to these traditional ones. In eastern India, people also drink lemon tea or lemon masala tea. Lemon tea simply contains hot tea with lemon juice and sugar. Masala lemon tea contains hot tea with roasted cumin seed powder, lemon juice, black salt and sugar, which gives it a tangy, spicy taste.

Adding a piece of ginger when brewing tea is a popular habit of Sri Lankans, who also use other types of spices such as cinnamon to sweeten the aroma. Other popular additives to tea by the tea-brewer or drinker include sugar, liquid honey or a solid Honey Drop, agave nectar, fruit jams, and mint. In China, sweetening tea was traditionally regarded as a feminine practice. In colder regions, such as Mongolia, Tibet and Nepal, butter is added to provide necessary calories. Tibetan butter tea contains rock salt and dre, a butter made from yak milk, which is churned vigorously in a cylindrical vessel closely resembling a butter churn. The same may be said for salt tea, which is popular in the Hindu Kush region of northern Pakistan.

Alcohol, such as whisky or brandy, may also be added to tea. The flavour of the tea can also be altered by pouring it from different heights, resulting in varying degrees of aeration. The art of high-altitude pouring is used principally by people in Northern Africa (e.g. Morocco, Algeria, Mauritania, Libya and Western Sahara), but also in West Africa (e.g. Guinea, Mali, Senegal) and can positively alter the flavour of the tea, but it is more likely a technique to cool the beverage destined to be consumed immediately. In certain cultures, the tea is given different names depending on the height from which it is poured. In Mali, gunpowder tea is served in series of three, starting with the highest oxidisation or strongest, unsweetened tea (cooked from fresh leaves), locally referred to as "strong like death", followed by a second serving, where the same tea leaves are boiled again with some sugar added ("pleasant as life"), and a third one, where the same tea leaves are boiled for the third time with yet more sugar added ("sweet as love"). Green tea is the central ingredient of a distinctly Malian custom, the "Grin", an informal social gathering that cuts across social and economic lines, starting in front of family compound gates in the afternoons and extending late into the night, and is widely popular in Bamako and other large urban areas.

In Southeast Asia, particularly in Singapore and Malaysia, the practice of pouring tea from a height has been refined further using black tea to which condensed milk is added, poured from a height from one cup to another several times in alternating fashion and in quick succession, to create a tea with entrapped air bubbles creating a frothy "head" in the cup. This beverage, teh tarik, literally, "pulled tea" (which has its origin as a hot Indian tea beverage), has a creamier taste than flat milk tea and is extremely popular in the region.

Tea pouring in Malaysia has been further developed into an art form in which a dance is done by people pouring tea from one container to another, which in any case takes skill and precision. The participants, each holding two containers, one full of tea, pour it from one to another. They stand in lines and squares and pour the tea into each other's pots. The dance must be choreographed to allow anyone who has both pots full to empty them and refill those of whoever has no tea at any one point.

ECONOMICS

Tea is the most popular manufactured drink consumed in the world, equaling all others – including coffee, chocolate, soft drinks, and alcohol – combined. Most tea consumed outside East Asia is produced on large plantations in the hilly regions of India and Sri Lanka, and is destined to be sold to large businesses. Opposite this large-scale industrial production are many small "gardens," sometimes minuscule plantations, that produce highly sought-after teas prized by gourmets. These teas are both rare and expensive, and can be compared to some of the most expensive wines in this respect.

India is the world's largest tea-drinking nation, although the per capita consumption of tea remains a modest 750 grams per person every year. Turkey, with 2.5 kg of tea consumed per person per year, is the world's greatest per capita consumer.

PRODUCTION

In 2003, world tea production was 3.21 million tonnes annually. In 2010, world tea production reached over 4.52 million tonnes after having increased by 5.7% between 2009 and 2010. Production rose by 3.1% between 2010 and

2011. The largest producers of tea are the People's Republic of China, India, Kenya, Sri Lanka, and Turkey. The following table shows the amount of tea production (in tonnes) by leading countries in recent years. Data are generated by the Food and Agriculture Organization of the United Nations as of February 2012.

Rank	Country	2008	2009	2010	2011
1	China	1,257,600	1,359,000	1,450,000	1,623,000
2	India	987,000	972,700	991,180	1,063,500
3	Kenya	345,800	314,100	399,000	377,912
4	Sri Lanka	318,700	290,000	282,300	327,500
5	Turkey	198,046	198,601	235,000	221,600
6	Vietnam	173,500	185,700	198,466	206,600
7	Iran	165,717	165,717	165,717	162,517
8	Indonesia	150,851	146,440	150,000	142,400
9	Argentina	80,142	71,715	88,574	96,572
10	Japan	96,500	86,000	85,000	82,100
Total	World	4,211,397	4,242,280	4,518,060	4,321,011

Labor and Consumer Safety Problems

Multiple recent reports have found that most Chinese and Indian teas contain residues of banned toxic pesticides. Tea production in Kenya, Malawi, Rwanda, Tanzania, and Uganda has been reported to make use of child labor according to the U.S. Department of Labor's List of Goods Produced by Child Labor or Forced Labor (a report on the worst forms of child labor).

Certification

Workers who pick and pack tea on plantations in developing countries can face harsh working conditions and may earn below the living wage. A number of bodies independently certify the production of tea. Tea from certified estates can be sold with a certification label on the pack. The most important certification schemes are Rainforest Alliance, Fairtrade, UTZ Certified, and Organic, which also certify other crops such as coffee, cocoa and fruit. Rainforest Alliance certified tea is sold by Unilever brands Lipton and PG Tips in Western Europe, Australia and the US. Fairtrade certified tea is sold by a large number of suppliers around the world. UTZ Certified announced a partnership in 2008 with Sara Lee brand Pickwick tea. Production of organic tea has risen since its introduction in 1990 at Rembeng, Kondoli Tea Estate, Assam. 6,000 tons of organic tea were sold in 1999. About 75% of organic tea production is sold in France, Germany, Japan, the United Kingdom, and the United States.

TRADE

According to the FAO in 2007, the largest importer of tea, by weight, was the Russian Federation, followed by the United Kingdom, Pakistan, and the

United States. Kenya, China, India and Sri Lanka were the largest exporters of tea in 2007 (with exports of: 374229, 292199, 193459 and 190203 tonnes respectively). The largest exporter of black tea is Kenya, largest producer (and consumer) India.

PACKAGING

TEA BAGS

In 1907, American tea merchant Thomas Sullivan began distributing samples of his tea in small bags of Chinese silk with a drawstring. Consumers noticed they could simply leave the tea in the bag and reuse it with fresh tea. However, the potential of this distribution/packaging method would not be fully realised until later on. During World War II, tea was rationed in the United Kingdom. In 1953 (after rationing in the UK ended), Tetley launched the tea bag to the UK and it was an immediate success.

The "pyramid tea bag" (or sachet) introduced by Lipton and PG Tips/ Scottish Blend in 1996, attempts to address one of the connoisseurs' arguments against paper tea bags by way of its three-dimensional tetrahedron shape, which allows more room for tea leaves to expand while steeping. However, some types of pyramid tea bags have been criticised as being environmentally unfriendly, since their synthetic material is not as biodegradable as loose tea leaves and paper tea bags.

LOOSE TEA

The tea leaves are packaged loosely in a canister, paper bag, or other container such as a tea chest. Some whole teas, such as rolled gunpowder tea leaves, which resist crumbling, are sometimes vacuum packed for freshness in aluminised packaging for storage and retail. The loose tea must be individually measured for use, allowing for flexibility and flavor control at the expense of convenience. Strainers, tea balls, tea presses, filtered teapots, and infusion bags prevent loose leaves from floating in the tea and over-brewing. A traditional method uses a three-piece lidded teacup called a gaiwan, the lid of which is tilted to decant the tea into a different cup for consumption.

COMPRESSED TEA

Compressed tea (such as Pu-erh) is produced for convenience in transport, storage, and ageing. It can usually be stored longer without spoilage than loose leaf tea. Compressed tea is prepared by loosening leaves from the cake using a small knife, and steeping the extracted pieces in water. During the Tang dynasty, as described by Lu Yu, compressed tea was ground into a powder, combined with hot water, and ladled into bowls, resulting in a "frothy" mixture. In the Song dynasty, the tea powder would instead be whisked with hot water in the bowl. Although no longer practiced in China today, the whisking method of preparing powdered tea was transmitted to Japan by Zen Buddhist monks, and is still used to prepare matcha in the Japanese tea ceremony. Compressed tea was the most popular form of tea in China during the Tang dynasty. By the beginning of the Ming dynasty, it had been displaced by loose leaf tea. It remains popular, however, in the Himalayan countries and Mongolian steppes. In Mongolia, tea bricks were ubiquitous enough to be used as a form of currency. Among Himalayan peoples, compressed tea is consumed by combining it with yak butter and salt to produce butter tea.

INSTANT TEA

"Instant tea", both hot and cold, is an alternative to the brewed products. Similar to freeze-dried instant coffee, but not requiring boiling water, instant tea was developed in the 1930s. Nestlé introduced the first commercial product in 1946, while Redi-Tea debuted instant iced tea in 1953.

Delicacy of flavour is sacrificed for convenience. Additives such as chai, vanilla, honey or fruit, are popular, as is powdered milk. During the Second World War British and Canadian soldiers were issued an instant tea known as 'Compo' in their Composite Ration Packs. These blocks of instant tea, powdered milk, and sugar were not always well received. As Royal Canadian Artillery Gunner, George C Blackburn observed:

- But, unquestionably, the feature of Compo rations destined to be remembered beyond all others is Compo tea...Directions say to "sprinkle powder on heated water and bring to the boil, stirring well, three heaped teaspoons to one pint of water."
- Every possible variation in the preparation of this tea was tried, but...it always ended up the same way. While still too hot to drink, it is a good-looking cup of strong tea. Even when it becomes just cool enough to be sipped gingerly, it is still a good-tasting cup of tea, if you like your tea strong and sweet. But let it cool enough to be quaffed and enjoyed, and your lips will be coated with a sticky scum that forms across the surface, which if left undisturbed will become a leathery membrane that can be wound around your finger and flipped away...

BOTTLED AND CANNED TEA

Canned tea is sold prepared and ready to drink. It was introduced in 1981 in Japan. In 1983, Swiss-based Bischofszell Food Ltd., was the first company to bottle ice tea on an industrial scale.

STORAGE

Storage conditions and type determine the shelf life of tea. Black tea's is greater than green's. Some, such as flower teas, may last only a month or so.

Others, such as pu-erh, improve with age. To remain fresh and prevent mold, tea needs to be stored away from heat, light, air, and moisture. Tea must be kept at room temperature in an air-tight container. Black tea in a bag within a sealed opaque canister may keep for two years. Green tea deteriorates more rapidly, usually in less than a year. Tightly rolled gunpowder tea leaves keep longer than the more open-leafed Chun Mee tea. Storage life for all teas can be extended by using desiccant or oxygen-absorbing packets, vacuum sealing, or refrigeration in air-tight containers (except green tea, where discrete use of refrigeration or freezing is recommended and temperature variation kept to a minimum).

3

Tea Culture

Tea culture is defined by the way tea is made and consumed, by the way the people interact with tea, and by the aesthetics surrounding tea drinking, it includes aspects of tea production, tea brewing, tea arts and ceremony, society, history, health, ethics, education, and communication and media issues. Tea is commonly consumed at social events, and many cultures have created intricate formal ceremonies for these events. Western examples of these are afternoon tea and the tea party. Tea ceremonies, with its roots in the Chinese tea culture, differ among eastern countries, such as the Japanese or Korean tea ceremony.

However, it may differ in preparation, such as in Tibet, where tea is commonly brewed with salt and butter. Tea plays an important role in some countries. The British Empire spread its own interpretation of tea to its dominions and colonies including regions that today comprise the states of India, Hong Kong, and Pakistan which had existing tea customs, as well as regions such as East Africa (modern day Kenya, Tanzania, and Uganda) and the Pacific (Australia, New Zealand) which did not have tea customs. Different regions favor different varieties of tea — black, green, or oolong — and use different flavourings, such as milk, sugar or herbs. The temperature and strength of the tea likewise varies widely.

BLACK DRINK

The black drink is a name for several kinds of ritual beverages brewed by Native Americans in the Southeastern United States. Traditional ceremonial people of the Caddo, Cherokee, Choctaw, Muscogee and some other Indigenous peoples of the Southeastern Woodlands use the black drink in purification ceremonies. It was occasionally known as white drink because of the association of the color white with peace leaders in some Native cultures in the Southeast.

The black drink is typically seen in Native cultures where it is present as a purifier that removes spiritual and physical contamination from the drinker, and as such is never taken casually. While the recipe can vary between tribes, and the full formula is not published or given to outsiders, a prominent ingredient in many formulas is the roasted leaves and stems of Ilex vomitoria native to the Atlantic and Gulf Coasts. The black drink usually contains emetic herbs, which are used with the express purpose of inducing vomiting. There is a degree of variation in its ritual use. It often involves vomiting, but not always. Vomiting is not induced by the mere consumption of a drink prepared with Ilex vomitoria. It was occasionally used by some groups as a stimulating, but not casual social drink for before council meetings.

PREPARATION

According to the ethnohistorical record, the yaupon leaves and branches used for the black drink were traditionally picked as close to the time of its planned consumption as possible. After picking they were lightly parched in a ceramic container over fire. The roasting increases the solubility in water of the caffeine, which is the same reason coffee beans are roasted. After browning, they were boiled in large containers of water until the liquid reached a dark brown or black color, giving it its name.

The liquid was then strained into containers to cool, until it was cool enough to not scald the skin, and drunk while still hot. Because caffeine is 30 times more soluble in boiling water than room temperature water, this heightened its effect. It was then consumed in a ritual manner. Its physiological effects are mainly those of massive doses of caffeine.

Three to six cups of strong coffee is equal to 0.5 to 1.0 grams of caffeine; the black drink could have delivered at least this much and possibly up to 3.0 to 6.0 grams of caffeine. Similar methods of production were adopted by European colonists for the production of a drink that often shared the same names with Native names for the black drink, but were used for different purposes.

While the general method of production is known, not all details of the preparation and ceremonial usage of the black drink are.The source of the emetic effect of black drink is not known and has been speculated upon by historians, archaeologists, and botanists. Some professionals belief it be caused by the addition of poisonous Button snakeroot. It has also been speculated that the emetic effect was caused by physically triggering the gag reflex or by the mere act of consuming vast quantities of a warm liquid.

Modern preparation and usage the black drink by Native Americans is less well documented. Online recipes for the black drink have been criticized by some Native Americans as potentially dangerous if not poisonous due those recipes leaving out key steps. The berries of the yaupon holly are poisonous. Kidney failure is one possible outcome of consuming beverages containing holly leaves. Those recipes have also been seen as violating Indigenous intellectual property and potentially leading to cultural appropriation. Some Native Americans warn that recipes found online or in books should not be assumed to be safe.

PRECONTACT USE

Archaeologists have demonstrated the use of various kinds of black drink among Native American groups stretching back far into antiquity, possibly dating to Late Archaic times. During the Hopewell period, the shell cups known from later black drink rituals become common in high status burials along with mortuary pottery and engraved stone and copper tablets. The significance of the shell cups may indicate the beginning of black drink ceremonialism. The fact that both the shells and the yaupon holly come from the same geographical location may indicate they were traded together in the Hopewell Interaction Sphere. The appearance of shell cups can be used as a virtual marker for the advent of Hopewell Culture in many instances. During the Mississippian culture period, the presence of items associated with the black drink ceremony had spread over most of the south, with many hundreds of examples from Etowah, Spiro, Moundville and Hiwassee Island.

Black drink at Cahokia

Archaeologists working at Cahokia, the largest Mississippian culture settlement located near the modern city of St. Louis, found distinctive and relatively rare pottery beakers dating from 1050 to 1250 CE. The beakers are small round pots with a handle on one side and a tiny lip on the opposing side. The surfaces of the unfired vessels was incised with motifs representing water and the underworld and resemble the whelk shells known to have been used for the consumption of the beverage during historic times. The inside of the vessels were found to be coated with a plant residue, which when tested was found to contain theobromine, caffeine and ursolic acid in the right proportions to have come from the Ilex vomitoria. The presence of the black drink in the Greater Cahokia area at this early date pushes back definitive use of the black drink by several centuries. The presence of the black drink hundreds of miles outside of its natural range on the East and Gulf coasts is evidence of a substantial trade network with the southeast, a trade that also involved sharks teeth and whelk shells.

Shell Cups

In historic accounts from the 16th and 17th century, black drink is usually imbibed in rituals using a cup made of marine shell. Three main species of marine shells have been identified as being used as cups for the black drink, lightning whelk, emperor helmet, and the horse conch. The most common was the lightning whelk, which has a left-handed or sinistral spiral. The left-handed spiral may have held religious significance because of its association with dance and ritual. The center columnella, which runs longitudinally down the shell, would be removed, and the rough edges sanded down to make a dipper like cup. The columnella would then be used as a pendant, a motif that shows up frequently in Southeastern Ceremonial Complex designs. In the archaeological record columnella pendants are usually found in conjunction with bi-lobed arrows, stone maces, earspools, and necklace beads(all of which are motifs identified with the falcon dancer/warrior/chunkey player mythological figure). Artifacts made from these marine shells have been found as far north as Wisconsin and as far west as Oklahoma. Several examples of cups from Moundville and Spiro have been found to have rings of black residue in the bottoms, suggesting they were used for black drink rituals. Many examples of shell cups found in Mississippian culture mounds are engraved with S.E.C.C. imagery. A few examples portray what is theorized to be black drink rituals, including what some anthropologists have interpreted as vomit issuing from the mouths of mythological beings.

POST CONTACT USE

A number of tribes across the Southeastern United States use a form of the black drink in their ceremonies. Muscogee Creeks, Cherokees, Choctaws, Ais, Guale, Chickasaws, Chitimacha, Timucua and others are documented users of a type of black drink. Although rituals vary amongst the different tribes, there are some common traits among them. Black drink is most commonly drunk by men, who also prepare it, but there are some accounts of its preparation and use by women among some groups. Removal of tribes to areas outside the natural range of Ilex vomitoria has been partially responsible for a decline in the preparation of the black drink among present Native Americans.

Ais

In 1696, Jonathan Dickinson witnessed the use of a beverage brewed from the Yaupon Holly among the Ais of Florida. Dickinson later learned that the Spanish called the plant casseena. The Ais parched the leaves in a pot, and then boiled them. The resulting liquid was then transferred to a large bowl using a gourd that had a long neck with a small hole at the top, and a 2-inch-wide hole in the side. On the occasion Dickinson witnessed, he estimated that there was nearly three gallons of the beverage in the bowl. After the liquid had cooled, the chief was presented with a conch-shell of the beverage. The chief threw part of it on the ground as a blessing and drank the rest. The chief's associates were then served in turn. Lower status men, women and children were not allowed to touch or taste the beverage. The chief and his associates sat drinking this brew, smoking and talking for most of the day. In the evening the bowl that had held the beverage was covered with a skin to make a drum. The Ais, accompanied by the drum and some rattles, sang and danced until the middle of the night.

Cherokee

Cherokee black drink is taken for purification at several traditional ceremonies. Made with emetics, the complete recipe is not shared with the public. The black drink induces vomiting for purification purposes. Other ritual medicinal beverages are also used in the ceremonies. Ilex vomitoria was not native to the traditional area controlled by the Cherokee before removal, but it was brought in to the area through trade and then transplanted.

Muscogean Peoples

Among the Muscogee the blank drink is called ássi. In the ceremonies of some cultures that use the drink, after its preparation it is passed out to the highest status person first, then the next highest status, and so forth. During each persons turn to drink, ritual songs may be sung Yahola. Its use was traditionally limited to only adult men. The ritual name Asi Yahola or Black Drink Singer is corrupted into English as Osceola).

Timucua

Among the Timucua, a type of black drink was called cacina by the Spanish. Preparation and consumption of the drink was strictly limited to the community council house. Women (other than an occasional female chief) were normally excluded from the council house except for activities such as dances, but did prepare the cacina. In 1678 a bedridden cacica was given permission to brew and consume cacina in her house, on the condition that no one else could be present while she did so.

CASSINA OR YAUPON TEA

After European contact with tribes in what is today the Southeastern United States, colonists began using the charred leaves of the yaupon holly to make a tea similar to the black drink, but without the ritual vomiting of it. Its use by colonists in Spanish Florida is documented as far back as 1615. An account from that year describes Spaniards as experiencing symptoms that would now be described as caffeine dependence due to daily consumption of what they called cacina or té del indio. The use of Ilex vomitoria by colonists for tea making and for medicinal uses in the Carolinas is documented by the early eighteenth century. In the English speaking colonies it was known variously as cassina, yaupon tea, Indian tea, Carolina tea, and Appalachian tea. It was commonly believed to be and used as a diuretic.

By the late 1700s, yaupon tea was described as being more commonly used in North Carolina at breakfast than tea made with Camellia sinensis. In addition to using it on their own, Spanish and British colonist often drank black drink when engaging in discussions and treaties with Natives. Its preparation by British and Spanish colonists was nearly identical to the method of preparation used by their Native neighbors. Its use by colonists in French Louisiana is speculated to have occurred, but lacks documentation other than one source describing its medicinal uses from 1716.

During the Civil War, yaupon tea was used as a substitute for coffee and tea throughout the South. Yaupon continued to be used in North Carolina for medicinal purposes and as a common drink until the late 1890s. At that point, its use was stigmatized because of its natural abundance as being a habit associated with rural poor people. By 1928 it was described as only being in common use by on Knotts Island, North Carolina. During the Interwar period the United States Department of Agriculture investigated the use of cassina tea as a substitute for coffee and tea. There were also a few attempts at the commercialization of cassina tea during that same period. By 1973 it was believed that cassina tea was only being served at the Pony Island Restaurant on Orcracoke Island, North Carolina. In the early 2000s yaupon tea began witnessing a resurgence in its popularity, and can now be purchased online and at several historical sites related to Native Americans.

BUBBLE TEA

Bubble tea (also known as pearl milk tea, or boba-milk tea) is a Taiwanese tea-based drink invented in Taichung in the 1980s. Most bubble tea recipes contain a tea base mixed/shaken with fruit or milk, to which chewy tapioca balls and/or fruit jellies are often added. The "bubble" refers to the foam created by shaking the tea, which itself is called "pàomò" tea, meaning "frothy/foamy" tea. Ice-blended versions are usually mixed with fruit or syrup, resulting in a slushy consistency.

There are many variants of the drinks, and many kinds of ingredients may be added. The most popular bubble drinks are bubble milk tea with tapioca and bubble milk green tea with tapioca.

Bubble teas are typically of two distinct types: fruit-flavored teas and milk teas. However, some shops offer hybrid "fruit milk teas". Most milk teas include powdered dairy or non-dairy creamers, but some shops also offer fresh milk as an alternative. Other varieties are 100% crushed-fruit smoothies with tapioca pearls and signature ice cream shakes made from local ice cream sources. Many American bubble tea vendors sell "milk smoothies", which are similar to bubble tea but do not contain any tea ingredients. Some small cafés offer sweetener substitutes, such as honey, agave, stevia, and aspartame, upon special request.

The oldest known bubble tea consisted of a mixture of hot Taiwanese black tea, small tapioca pearls, condensed milk, and syrup or honey. Many variations were created, the most common of which is served cold rather than hot. The tea type is frequently replaced. First was bubble green tea, which uses jasmine-infused green tea instead of black tea. Big tapioca pearls were adapted and quickly replaced the small pearls. Peach or plum flavoring appeared, then more fruit flavors were added until, in some variations, the tea was removed entirely in favor of real fruit. These fruit versions sometimes contain colored pearls (and/or "jelly cubes" as in the related drink taho), the color chosen to match whatever fruit juice is used. Flavors may be added in the form of powder, fruit juice, pulp, or syrup to hot black or green tea, which is then shaken in a cocktail shaker or mixed with ice in a blender. Cooked tapioca pearls and other mix-ins (such as vanilla extract, honey, syrup, and sugar) are added at the end.

Today, one can find shops entirely devoted to bubble tea, similar to the juice bars of the early 1990s. Some cafés use plastic dome-shaped lids, while other bubble tea bars serve it using a machine to seal the top of the cup with plastic cellophane. The latter method allows the tea to be shaken in the serving cup and makes it spill-free until one is ready to drink it. The cellophane is then pierced with an oversized straw large enough to allow the pearls to pass through. Today, in Taiwan, it's more common for people to refer to the drink as "pearl milk tea" ("zhen zhû nai chá", or "zhen nai" for short). "Pearl milk tea" is also used by English speakers and overseas Chinese and Taiwanese speakers, but it is usually called "bubble tea" or "boba tea" by English speakers, with the former seemingly more common in locations with less Chinese influence.

VARIANTS

Each of the ingredients of bubble tea can have many variations depending on the tea store. Typically, different types of black tea, green tea, or even coffee form the basis of this beverage. The most common black tea varieties are oolong and Earl Grey, while jasmine green tea is a mainstay at almost all tea stores. Another variation called yuanyang originated in Hong Kong and consists of half black tea and half coffee. Some people add milk to the drink. Decaffeinated versions of teas are sometimes available when the tea house freshly brews the tea base.

The milk in bubble tea is optional, though many tea stores use it. Some cafés use a non-dairy creamer milk substitute instead of milk because many East Asians are lactose intolerant and because it is cheaper and easier to store and use than perishable milk. In Western countries, soy milk options are widely available for those who avoid dairy products. This adds a distinct flavor and consistency to the drink.

Different flavorings can be added to bubble tea. Some widely available fruit flavors include strawberry, green apple, passion fruit, mango, lemon, watermelon, grape, lychee, peach, pineapple, cantaloupe, honeydew, banana, avocado, coconut, kiwifruit, and jackfruit. Other popular non-fruit flavors include taro, pudding, chocolate, coffee, mocha, barley, sesame, almond, ginger, lavender, rose, caramel, and violet. Some of the sour fruit flavors are available in bubble tea without milk only as the acidity tends to curdle the milk.

Other varieties of the bubble tea drink can include blended drinks. Many stores in the US provide a list of choices to choose from. Some may include coffee-blended drinks or even smoothies. Tapioca balls are the prevailing chewy tidbits in bubble tea, but a wide range of other options can be used to add similar texture to the drink. Green pearls have a small hint of green tea flavor and are chewier than the traditional tapioca balls. Jelly is also used in small cubes, stars, or rectangular strips, with flavors such as coconut jelly, konjac, lychee, grass, mango, and green tea often available at some shops. Rainbow, a fruit mix of konjac, has a pliant, almost crispy consistency. Azuki bean or mung bean paste, also typical toppings for Taiwanese shaved ice desserts, give the drinks an added subtle flavor as well as texture. Aloe, egg pudding (custard pudding), sago, and taro balls can also be found in most tea houses.

Due to its popularity, single-serving packets of black tea (with powdered milk and sugar included) are available as "Instant Boba Milk Tea" in some places. Bubble tea cafés will also frequently serve drinks without coffee or tea in them. The base for these drinks is flavoring blended with ice, often called Snow Bubble. All mix-ins that can be added to the bubble tea can also be added to these slushie-like drinks. One drawback to them is that the coldness of the iced drink may cause the tapioca balls to harden, making them difficult to suck up through a straw and chew. To prevent this from happening, these slushies must be consumed more quickly than bubble tea.

Occasionally, nata de coco is used in mass-produced bubble tea drinks as a healthier alternative to tapioca starch. Nata de coco is high in dietary fiber and low in cholesterol and fat. The nata de coco is sliced into thin strips to make it easier to pass through a straw.

TYPES

- Pàomò hóngchá: "foam red tea", by direct translation, is the drink that is more appropriate for the more literal name of "bubble tea"; however, the English name, foam tea, is not used much throughout Asia. Consequently, in non-Chinese-speaking Asian countries, "bubble tea" is commonly used to refer to this drink. There is no tapioca in this particular drink. To create this, vendors mix hot or warm tea (in this case, black tea) with syrup or sugar and ice cubes into a cocktail shaker. Then they would shake the shaker either by hand or by machine before it is served. The resulting tea would be covered by a layer of foam or froth and the tea would have a light foamy feel to the taste.
- Pàomò naichá: "foam milk tea". One of the many variants that is prepared the same way as the "foam red tea", well-shaken before serving.
- Zhenzhû naichá: "pearl milk tea", or more commonly referred to as bubble tea by most English speakers and overseas Chinese speakers. The "pearl" name originally referred to the small 1/12" tapioca pearls added to the drink. Though most modern vendors serve only the bigger 1/4" pearls, they still use "pearl tea" as the name.

- Bôbà naichá: "bubble milk tea" and also commonly referred to as boba tea by English speakers and Asian Americans. The name refers to the variant with the bigger, 1/4" tapioca pearls.
- Hei zhenzhû naichá: "black pearl milk tea". Since the bigger 1/4" tapioca pearls are separately sold as "black pearls" in markets, this name was the logical first choice and is more popular with the consumers of the drink. "Boba" is an alternative name that is less commonly used these days.
- Chá zhenzhû): " tea pearl" .
- Pào pào chá: used interchangeably with refer to "bubble tea" in Singapore.

HEALTH CONCERNS

Tapioca pearls, milk powder, and juice syrups have in the past been found to contain banned chemical additives. In May 2011, a food scandal broke out in Taiwan where DEHP (a chemical plasticizer and potential carcinogen used to make plastic) was found as a stabilizer in drinks and juice syrups. Some of these products may have been exported and used in bubble tea shops around the world. DEHP can affect hormone balances. In June 2011, the Health Minister of Malaysia, Liow Tiong Lai, instructed companies selling "Strawberry Syrup", a material used in some bubble teas, to stop selling them after chemical tests showed they were tainted with a carcinogen identified as DEHP.

In August 2012, scientists from the Technical University of Aachen in Germany analyzed bubble tea samples within a research project to look for allergenic substances. The result indicated that the products contain styrene, acetophenone and brominated substances. The report was published by German newspaper Rheinische Post and caused Taiwan's representative office in Germany to issue a statement, saying food items in Taiwan are monitored. Taiwan's Food and Drug Administration confirmed in September that in a second round of tests conducted by German authorities, Taiwanese bubble tea was found to be free of cancer-causing chemicals. The products were also found to contain no excessive levels of heavy-metal contaminants or other health-threatening agents.

In May 2013 the Taiwan Food and Drug Administration issued an alert on the detection of maleic acid, an unapproved food additive, in some food products, including tapioca pearls. The Agri-Food & Veterinary Authority of Singapore conducted its own tests and found additional brands of tapioca pearls and some other starch-based products sold in Singapore were similarly affected.

In April 2015 investigation for cleaning a pitcher containing the alleged poisonous milk tea at the store namely ErgoCHA in Sampaloc, Manila Philippines denied the allegation of their helper that he brought a chemical to the store the night before the alleged poisoning happened. However, preliminary findings of the Department of Health and Food and Drug Administration did not find any toxic substance in the sample milk tea. The DOH and FDA are still conducting further tests to determine what caused the deaths. In a news report, ABS-CBN News told that the pitcher containing the alleged poisonous milk tea might be containing a cyanide.

As of May 10, 2015, ABS-CBN's news program TV Patrol reported that the victims of the milk tea poisoning at ErgoCha in Sampaloc, Manila was contaminated by a cleaning agent, oxalic acid. Although, investigation is still on-going for further information about the said incident.

HISTORY

The most accredited story about the creation of bubble tea is the Chun Shui Tang Teahouse in Taichung, Taiwan. The founder of the teahouse, Liu Han-Chieh, observed how the Japanese served cold coffee while on a visit in the 1980s, and applied this method to tea. The new style of serving tea propelled his business, and multiple chains were established. This expansion would be the stepping stone for the rapid expansion of bubble tea. The actual creator of bubble tea is Lin Hsiu Hui, the teahouse's product development manager, who poured a sweetened pudding with tapioca balls into the iced tea drink during a meeting in 1988. The beverage was well received by the people at the meeting, leading to its inclusion on the menu. It ultimately became the franchise's top-selling product. An alternative origin is the Hanlin teahouse in Tainan, Taiwan, owned by Tu Tsong-he. He made tea using traditional white tapioca, which has the appearance of pearls, supposedly resulting in the so-called "pearl tea". Shortly after, Hanlin changed the white tapioca balls to the black version that is seen most today.

The drink became popular in most parts of East and Southeast Asia during the 1990s. Famous chain-stores that sell boba teas are Gong Cha, ShareTea, and Coco The drink is well received by global consumers from Canada and USA, specifically around areas with high Asian demographics.

BUTTER TEA

Butter tea, also known as po cha, cha süma, Mandarin Chinese or gur gur in local Ladakhi terms, is a drink of the people in the Himalayan regions of Nepal, India and Bhutan and most famously, Tibet. It is also consumed in Bhutan and by Buddhist minorities in India. Traditionally, it is made from tea leaves, yak butter, water, and salt, however, given its wider availability and reduced cost, butter made from cow's milk is increasingly used. Drinking butter tea is a regular part of Tibetan life. Before work, a Tibetan will typically enjoy several bowlfuls of this beverage, and it is always served to guests. Nomads are said to often drink up to 40 cups of it a day. Since butter is the main ingredient, butter tea provides plenty of caloric energy and is particularly suited to high altitudes. The butter may also help prevent chapped lips.

According to the Tibetan custom, butter tea is drunk in separate sips, and after each sip the host refills the bowl to the brim. Thus, the guest never drains his bowl; rather, it is constantly topped up. If the visitor does not wish to drink, the best thing to do is leave the tea untouched until the time comes to leave and then drain the bowl. In this way etiquette is observed and the host will not be offended. Butter tea is also used for eating tsampa by pouring onto it, or dipping the tsampa into it, and mixing well. The concentrate, produced by repeatedly boiling tea leaves, will keep for several days, and is commonly used in towns. The tea is then combined with salt and butter in a special tea churn and churned vigorously before serving hot. Now an electric blender is often used.

HISTORY

While evidence of tea has been found in Tibet from before the 10th century, it did not reach its nearly universal status until about the 13th century, the time of the Sakya hierarchy and the Phagmodu kings. By the start of the rule by the Dalai Lama, tea had become a government monopoly. While it is now officially allowed to be sold by anyone, it still is mostly sold by government officials.

PREPARATION

The highest quality tea is made by boiling the tea leaves in water for half a day, achieving a dark brown color. It is then skimmed, and poured into a cylinder with fresh yak butter and salt which is then shaken. The result is a purplish liquid that is about the thickness of a stew or thick oil. It is then poured into clay tea-pots, or jars, that resemble Japanese teapots. Another method is to boil water, and add handfuls of the tea into the water, which is allowed to steep until it turns almost black. Salt is then added, along with a little soda if wanted. The tea is then strained through a horse-hair or reed colander into a wooden butter churn, and a large lump of butter is added. This is then churned until the tea reaches the proper consistency and transferred to copper pots that sit on a brazier to keep them warm. When a churn is not available, a wooden bowl and rapid stirring will suffice. Nowadays, when tea leaves, yak butter and wooden butter churn is not available, people often make butter tea using tea bags, different types of butter available in the market and a blender to churn.

CHA CHAAN TENG

A cha chaan teng meaning tea restaurant, is commonly found in Greater China, including particularly Hong Kong, Macau and parts of Guangdong. They are known for eclectic and affordable menus, which include many dishes from

Hong Kong cuisine and Hong Kong-style Western cuisine. Since the mass migration of Hong Kong people in the 1980s they are also commonplace in many Western countries like Australia, Canada, the United Kingdom and the United States, particularly in the Chinatown areas of many major cities.In the early Hong Kong, only high-class restaurants provided western food and most of them did not serve local people. At that time, people saw western food as a luxury item. After the Second World War, Hong Kong culture was influenced by British culture. Hong Kong people started to like drinking tea and eating cakes. Therefore, some of the Hong Kong people set up the Cha Chaan Teng and their target audience was local people.

Providing different kinds of Canto-Western Cuisine and drinks with very low price led to them being regarded as "cheap western food" or "Soy sauce western food". In recent years, the management of Cha Chaan Teng began to change in co-ordination with the development of Hong Kong economy and society. During the 1997 Asian Financial Crisis, Cha Chaan Teng became much more popular in Hong Kong as they still provided the cheapest food for the public. In April 2007, one of the Hong Kong political officers suggested that Cha Chaan Teng be listed in the UNESCO Intangible Cultural Heritage Lists, because of its important role in Hong Kong society.

Cha chaan teng establishments provide tea called "clear tea", to customers as soon as they are seated. Some patrons use the hot tea to wash their utensils. The name, literally "tea restaurant", serves to distinguish itself from Western restaurants that provide water to customers instead of tea. The "tea" in the name refers to the inexpensive black tea, not the traditional Chinese tea served in traditional dim sum restaurants and teahouses. Moreover, some cha chaan tengs prefer the use of the word "café" in their names.

The "tea" may also refer to tea drinks, such as the Hong Kong-style milk tea and iced lemon tea, which are very popular in cha chaan tengs. The older generations in Hong Kong use yum sai cha, when dining in these restaurants in contrast with yum cha.

CULTURE

- Fast service and high efficiency: Usually, tea restaurants have high efficiency, with each customer averagely spending 10–20 minutes to finish a meal. Customers typically receive their dishes after five minutes. The waiters take the order with their left hand and pass the dishes with their right hand. This embodies Hong Kong's hectic lifestyle. In rush hour, it is common for a lot of people to queue outside the restaurants.
- Long working hours: Cha Chaan Teng workers may work 07:00 to 23:00, some even working a night shift. Long working hours are characteristic of Hong Kong.

- Trend: Because of the limited lands and expensive rent, Cha Chaan Tengs are gradually being replaced by chain restaurants, such as Café de Coral, Maxim's and Fairwood. As chain restaurants dominate the market, Hong Kong's Cha Chaan Teng culture is disappearing.

MENUS

Cha chaan teng serves a wide range of food, from steak to wonton noodles to curry to sandwiches, e.g. Hong Kong-style French toast. Both fast food and à-la-carte dishes are available. A big cha chaan teng often consists of three cooking places: a "water bar" which makes drinks, toast/sandwiches and instant noodles, a "noodle stall" which prepares Chiuchow-style noodles , and a kitchen for producing rice plates and other more expensive dishes. The invention of drinks like yuanyang, Iced tea with Lemon and Coca-Cola with Lemon are often credited culturally to this style of restaurant.

FOOD AND DRINKS

Drinks:

- Coffee (Two types exist: instant and in powder form, in which the latter is more prominent)
- Black Coffee—Hong Kong people usually called them jaai fea(lit. "Vegetarian Coffee" / "just coffee", emphasising on its plain texture) or "fei sha jaau naai" (lit. "get rid of sand and milk", i.e. coffee without sugar and milk)
- Hong Kong-style milk tea—A drink highly popular in Hong Kong and its standard is judged by its aroma, smoothness and concentration. The tea is soaked in an iron container for several hours to let the flavour come out. If the customer wants condensed milk instead of normal milk, people will often say "cha zou"
- Yuanyang—A mixture of coffee and tea, originated in Hong Kong. According to traditional Chinese Medicine, coffee and tea is "hot" and "cold" in nature respectively. A mixing of both thus then yield the best combination for beverage.
- Black and White Yuanyang— A mixture of Ovaltine and Horlicks, originated in Hong Kong.
- Horlicks
- Ovaltine
- Lemon Tea
- Lemonade—Often goes without sugar or syrup.
- Lemon with Coffee
- Lemon with Ribena
- Lemon with Honey—Often, to reduce cost, the honey is substituted with "Watercress Honey"

- Soft drinks—Coca Cola, 7-up, Fanta and Cream Soda are some common selections.

Adding ice in a drink may cost an extra fee. Some people simply ask for a glass of ice.

Snacks:

- Toast— This includes toast with condensed milk and butter/peanut butter, toast with jam and butter, toast with butter in a sliced form, and French Toast
- Sandwich— Sandwiches found in Cha Chaan Tengs usually include egg, ham, corned beef or a mixture of any as a filling. Club Sandwiches are also very common. A difference between the sandwiches found in Cha Chaan Tengs, and other eateries would be that only white bread is used, the customer has the option of omitting the crust of the bread, and requesting that the bread be toasted before making their sandwich.
- Egg tart
- Hong Kong-style buns— Includes [pineapple buns] (with or without a slab of butter inside), cocktail buns, Satay beef buns, barbecued pork buns, etc.
- Spring rolls

Fried dishes:

- Various Fried rice and noodles dishes
- Hong Kong-style spaghetti bolognese
- Fried instant noodles
- Gon chaau ngau ho — "dry fried beef and rice noodles" is a common Cantonese dish, made from stir-fried beef, he fen , dark soya sauce, and bean sprouts.

Soup dishes:

- Instant or u-don noodles in soup
- Macaroni in soup
- Soup noodles with fish balls, wontons and the like

Miscellaneous Dishes:

- Barbecued Pork (Char siu) — Not found in all Cha Chaan Tengs.
- Congee and yau ja gwei, a Chinese fritter.

SET MEALS

A feature found in Cha Chaan Tengs are set meals. There are various sets available throughout the day for breakfast, lunch, afternoon tea and dinner.

The lunch and dinner sets usually include a soup and a drink. Generally there is an additional HK$2 charge for cold drinks. Sometimes an additional HK$1 is charged for toasted bread.

Other sets include:

- "Nutritious set" – It comes with milk and other nutritional food
- "Constant set"– Provided all day long, hence the name (it usually consists of a main course, omelette, wheat foodstuff, white bread with butter and a drink) The wheat food stuff always comes with different choices such as spiced pork cubes, salted vegetable with sliced pork or luncheon meat etc.
- "Fast set"– Immediately served (usually rice with sausages/ ham/ fried eggs with gravy)
- "Special set"– Chef's recommendation

TABLES AND SEATS

Generally, the tables in Cha Chaan Tengs are square for 4 people, or round for 6 to 8 people. For each table, there is a piece of glass that covers the top and some menus are placed between the table and glass. During lunch or dinner, customers are sometimes requested to "daap toi". This means to have them sit with other strangers at the same table to save space, get waiting people seats faster, and give customers in a hurry a spot.

OTHER INTERESTING FACTS

- Most Cha Chaan Tengs in Hong Kong and Shenzhen would offer a cup of tea (the staff would call them "Gaat Sui", literally "empty water") to each customer after they are seated. The tea is bland and Hong Kong people would usually use it to clean the eating utensils. The utensils may be provided in the following two ways: either the waiter provides the utensils required after taking order (such as fork, knife and spoon for rice with steaks / fillets on plate, and chopsticks for soup / fried noodles), or placing all the utensils in a circular container placed on the table. Chopsticks would be separated from fork, knife and spoon. The hygienic condition of some of the Cha Chaan Teng might not be satisfactory, so the utensils would be washed by the cup of tea before using them. Some newer generations of Cha Chaan Teng would place everything including napkins into a drawer under the table to save space.
- Before 2007, most Cha Chaan Tengs allowed people to smoke, and some waiters would even smoke when working. This was a problem to people who did not like the smoke when eating. Starting from 1 January 2007, Hong Kong Law prohibited everyone from smoking within the indoor premises of restaurants.
- Usually, spring onions are sprinkled on top of congee and soup noodles. One may ask for to omit those spring onions.
- Two methods of frying noodles such as beef chow fun are wet frying

The former technique means to fry with a thick sauce, while the latter means to fry with only soy sauce.

- Much of the plasticware found on the table is provided by beverage companies, which is a form of advertising. This plasticware includes containers holding toothpicks, plastic menu holders, etc. Brands like Ovaltine, Horlicks and Ribena are the usual providers.
- To minimise costs, Cha Chaan Tengs will usually not have utensils that bear their own brand name. As a result, the patterns found on their utensils are always the same. These utensils can easily be purchased in supermarkets and various department stores.
- Tiles are placed along the walls and floors for easy cleaning.

VARIATIONS

Other kinds of local restaurant related to cha chaan teng in Hong Kong include chaan sutt, bing sutt, and bing teng, which a provide a lighter and more limited selection of food than cha chaan teng. In the old days, these eateries only sold different types of "ice", sandwiches and pasta but no rice plates. However, some of the restaurants bearing these titles today ignore the tradition, and provide all kinds of rice plates and even wonton noodles. Original chaan sutts, bing sutts and bing tengs, which can be regarded as the prototype of cha chaan tengs, are now scarce in Hong Kong. In June 2009, Hong Kong retail design store G.O.D. collaborated with Starbucks and created a store with a "Bing Sutt Corner" at their store on Duddell Street. It is a concept that fuses the retro Hong Kong teahouse, style with the contemporary look of a coffeehouse.

IN MEDIA

- The similarities between the different set meals were satirised of by My life as McDull, a McDull movie.
- An important part of Hong Kong culture, cha chaan teng is featured in many Hong Kong movies and TV dramas, including the popular sitcom Virtues of Harmony. The TVB-made soap opera tells the story of a family who runs a cha chaan teng, usually boasting the egg tart and "silk-stocking milk tea" produced by them. Stephen Chow also played a cha chaan teng waiter in the 1998-comedy Lucky Guy and a cha chaan teng meal-delivery-boy in the King of Comedy .
- Some beverage companies put the term cha chaan teng on their products, such as "cha chaan teng milk tea" and "cha chaan teng lemon tea".
- On 19 December 2007, lawmaker Choy So Yuk proposed during a Legislative Council session that Hong Kong's cha chaan teng be recognised and put up to UNESCO as an "intangible cultural heritage

of humanity". The proposal came about after a recent Hong Kong poll found that seven out of ten people believe the cafes deserve a UNESCO cultural listing.

CHASHITSU

In Japanese tradition, architectural spaces designed to be used for tea ceremony gatherings are known as chashitsu. The architectural style that developed for chashitsu is referred to as the sukiya style (sukiya-zukuri), and the term sukiya may be used as a synonym for chashitsu. Related Japanese terms are chaseki, broadly meaning "place for tea," and implying any sort of space where people are seated to participate in tea ceremony, and chabana, "tea flowers", the style of flower arrangement associated with the tea ceremony. Typical features of chashitsu are shôji windows and sliding doors made of wooden lattice covered in a translucent Japanese paper; tatami mat floors; a tokonoma alcove; and simple, subdued colours and style. The ideal floor size of a chashitsu is 4.5 tatami mats.

TEA HOUSES VERSUS TEA ROOMS

In Japanese, free-standing structures specifically designed for exclusive tea ceremony use, as well as individual rooms intended for tea ceremony, are both referred to as chashitsu. The term may be used to indicate the tea room itself where the guests are received, or that room and its attached facilities, even extending to the roji garden path leading to it. In English, a distinction is often made between free-standing structures for tea, referred to as tea houses, and rooms used for tea ceremony incorporated within other structures.

Tea houses are usually small, simple wooden buildings. They are located in the gardens or grounds of private homes. Other common sites are the grounds of temples, museums, and parks. The smallest tea house will have two rooms: the main room where the host and guests gather and tea is served, and a mizuya, where the host prepares the sweets and equipment. The entire structure may have a total floor area of only three tatami mats.

Very large tea houses may have several tea rooms of different sizes; a large, well-equipped mizuya resembling a modern kitchen; a large waiting room for guests; a welcoming area where guests are greeted and can remove and store their shoes; separate toilets for men and women; a changing room; a storage room; and possibly several anterooms as well as a garden with a roji path, an outdoor waiting area for guests and one or more privies.

Tea rooms are purpose-built spaces for holding tea gatherings. They may be located within larger tea houses, or within private homes or other structures not intended for tea ceremony. A tea room may have a floor area as small as 1.75 tatami mats (one full tatami mat for the guests plus a tatami mat called a daime, approximately 3/4 the length of a full tatami mat, for the portable brazier or sunken hearth (ro) to be situated and the host to sit and prepare the tea); or as large as 10 tatami mats or more; 4.5 mats is generally considered the ideal in modern tea rooms. A tea room will usually contain a tokonoma and a sunken hearth for preparing tea in the winter.

HISTORY OF CHASHITSU

The term chashitsu came into use after the start of the Edo period . In earlier times, various terms were used for spaces used for tea ceremony, such as chanoyu zashiki, sukiya (place for poetically-inclined aesthetic pursuits) such as chanoyu), and kakoi.

According to Japanese historian Moriya Takeshi in his article "The Mountain Dwelling Within the City," the ideal of wabi-style tea ceremony had its roots in the urban society of the Muromachi period , and took form in the tea houses that townspeople built at their residences and which affected the appearance of thatched huts in mountain villages. Before this, tea ceremony was generally enjoyed in rooms built in the shoin-zukuri architectural style, a style frequently employed in tea rooms built today. Tea houses first appeared in the Sengoku period (mid-15th century to early 17th century), a time in which the central government had almost no practical power, the country was in chaos, and wars and uprisings were commonplace. Seeking to reclaim Japan, samurai were busy acquiring and defending territories, promoting trade and overseeing the output of farms, mills and mines as de facto rulers, and many of the poor were eager to seek the salvation of the afterlife as taught by Zen Buddhism. Tea houses were built mostly by Zen monks or by daimyo, samurai, and merchants who practiced tea ceremony. They sought simplicity and tranquility - central tenets of Zen philosophy. The acknowledgment of simplicity and plainness, which is a central motivation of the tea house, continued to remain as a distinct Japanese tradition in the later periods.

AN IDEAL TEA HOUSE

The ideal free-standing tea house is surrounded by a small garden having a path leading to the tea room. This garden is called roji or "dewy ground." Along the path is a waiting bench for guests and a privy. There is a stone water-basin near the tea house, where the guests rinse their hands and mouths before entering the tea room through a small, square door called nijiriguchi, or "crawling-in entrance," which requires bending low to pass through and symbolically separates the small, simple, quiet inside from the crowded, overwhelming outside world. The nijiriguchi leads directly into the tea room.

The tea room has a low ceiling and no furniture: the guests and host sit seiza-style on the floor. All materials used are intentionally simple and rustic. Besides the guests' entrance, there may be several more entrances; at minimum there is an entrance for the host known as the sadôguchi, which allows access to the mizuya. Windows are generally small and covered with shôji, which allows natural light to filter in. The windows are not intended to provide a view to the outside, which would detract from the participants' concentration. There is a sunken hearth located in the tatami adjacent to the host's tatami, for use in the cold months; this hearth is covered with a plain tatami and is not visible in the warm months, when a portable brazier is used instead.

There will be a tokonoma holding a scroll of calligraphy or brush painting, and perhaps a small, simple, flower arrangement called a chabana, but no other decoration.

TEA ROOM LAYOUTS

Chashitsu are broadly classified according to whether they have a floor area larger or smaller than 4.5 tatami, a differentiation which affects the style of ceremony that can be conducted, the specific type of equipment and decoration that can be used, the placement of various architectural features and the hearth, and the number of guests that can be accommodated. Chashitsu which are larger than 4.5 mats are called hiroma , while those that are smaller are called koma . Hiroma often are shoin style rooms, and for the most part are not limited to use for chanoyu.

Other factors that affect the tea room are architectural constraints such as the location of windows, entrances, the sunken hearth and the tokonoma, particularly when the tea room is not located within a purpose-built structure. Not all tea rooms can be made to conform to the ideal seen in the image.

TYPICAL NAMES FOR CHASHITSU

It is usual for chashitsu to be given a name after their owners or benefactors. Names usually include the character for "hut", "hall," or "arbour," and reflect the spirit of rustic simplicity of the tea ceremony and the teachings of Zen Buddhism. Characteristic names include:

- Mugai-an
- Mokurai-an
- Tôkyû-dô
- Shô-an
- Ichimoku-an
- Rokusô-an
- Bôji-tei

CHATIME

Chatime is a Taiwanese global franchise teahouse chain. It operates 1,002 outlets in 26 countries. In 2006, it opened its first store outside of Taiwan in California, and it has since expanded to Malaysia, Canada, the Philippines, Cambodia, Thailand, Australia, New Zealand, India, Dubai, China, Japan and

Korea, among others. Its expansion and growth model is through franchising. Its parent company, La Kaffa International Co Ltd debuted on the Emerging Stock Market in December 2012 at NT$168 per share. It is expected to list on Taiwan's Gre Tai Securities Market in the second half of 2013. It is the only public-listed bubble tea company in Taiwan.

Henry Wang Yao-Hui founded Chatime in 2005 in Hsinchu, Taiwan, under the parent company, La Kaffa Coffee. He modernised traditional tea making culture by producing tea machines that could make quality fresh teas that were consistent, using a variety of flavors, fruits and toppings, in order to stand out from the many competitors in the market. The company has its own tea plantation in Nantou, Taiwan.

The company opened its first outlet outside of Taiwan in California in 2006, followed by Australia in 2009, where it currently has over 40 stores. This was followed by Malaysia in 2010. Its Malaysian operations currently account for just under 50% of its global revenue. In 2012, the brand opened its first store in India, a traditional tea drinking country, in New Delhi. After its two stores in New York, its third store in the United States opened at Mall of America in March 2012. The Chatime store in London is located on Old Compton Street. In the Philippines, it planned to open 50 stores by 2014, while in Malaysia, it opened its 100th store in 2013, with plans to open up to 250 stores within a three to five-year period.

LISTING

Chatime's parent company, La Kaffa International Co. Ltd., debuted on Taiwan's Emerging Stock Market in December 2012 at NT$168 per share under the stock code 2732. It is the only public-listed bubble tea company in Taiwan.

PRODUCTS

The brand's best-selling drink is its signature Chatime "pearl" milk tea. The brand has a variety of drinks under categories such as milk tea, fresh tea, fruit tea, "oriental pop tea", "QQ" jelly, "mousse", juices, smoothies and coffee, with stores carrying anywhere between 30 to over 90 flavours. Its products are available with less sugar. Typically, Chatime products come with a variety of toppings such as "pearls", while "QQ" jelly is a mixture of "pearls" and nata de coco. Some products are only available in certain countries. For example, in Malaysia, Chatime introduced three flavours in collaboration with Horlicks from GlaxoSmithKline.

COLLABORATIONS

In 2012, AirAsia announced that it would offer Chatime on its flights and claimed that is would be the "world's first and only in-flight bubble tea". The airline said that two speciality bubble tea drinks, Roasted Milk Tea with Red Bean and Mango Green Tea with Rainbow Jelly, were available on flights operated by Malaysian AirAsia with the airline code AK to 48 destinations, which include India, Hong Kong, Thailand and the Philippines. The collaboration was organised by Chatime Malaysia's CEO, Bryan Loo. Chatime became the first beverage company in Malaysia to host an official account on WeChat. To mark the collaboration, Chartime gave away 1 million cups of drinks worth over RM6 million/US$2 million through digital coupons made available via the WeChat platform. Other collaborations include partnerships with McDonald's, Horlicks from GlaxoSmithKline, and biscuit brand Julie's.

BRAND AMBASSADORS

Chatime announced in 2012 that award-winning Taiwan actress Ariel Lin would serve as its brand ambassador. The following year, the brand announced that Crystal Lee had become the lifestyle tea provider's first "Chatime Friend". The announcement was made soon after Lee became the youngest actress ever to have won the award for Best Actress at the 16th Shanghai International Film Festival.

CREAM TEA

A cream tea (also known as a Devonshire tea, Devon cream tea or Cornish cream tea) is a form of afternoon tea light meal, consisting of tea taken with a combination of scones, clotted cream, and jam. Traditionally a speciality of Devon and Cornwall, cream teas are offered for sale in tea rooms in those two counties, as well as in other parts of England, and elsewhere in the Commonwealth. The name "Devonshire tea" comes from the county of Devon in England. The exact origin of "cream tea" is disputed, although there is evidence to suggest that the tradition of eating bread with cream and jam already existed at Tavistock Abbey in Devon in the 11th century.

VARIATIONS

There are regional variations as to how a cream tea should preferably be eaten. The Devonshire method is to split the scone in two, cover each half with clotted cream, and then add strawberry jam on top. Traditionally it is important that the scones be warm , and that clotted cream and strawberry jam, rather than any other variety, are used. Butter is generally not included, and the tea should be served with milk. In Cornwall, the cream tea was traditionally served with a "Cornish split", a type of slightly sweet white bread roll, rather than a scone. It is now rare to find this available commercially, except in Cornwall; splits are still used by many Cornish families in their own homes. The warm roll is first buttered, then spread with strawberry jam, and finally topped with a spoonful of clotted cream. Scones are rarely buttered in commercially available teas. Where clotted cream is not available, whipped cream is sometimes used as a substitute. Another variation to a cream tea is called "Thunder and Lightning", which consists of a round of bread, topped with clotted cream and golden syrup, honey, or treacle.

PROTECTED STATUS

In May 2010, a campaign was launched at the Devon County Show to have the name "Devon cream tea" protected within the European Union under Protected Designation of Origin rules. The campaign was launched following discussion on BBC Radio Devon.

DIM SUM

Dim sum is a style of Cantonese cuisine prepared as small bite-sized portions of food traditionally served in small steamer baskets or on small plates. Dim sum is also well known for the unique way it is served in some restaurants, whereby fully cooked and ready-to-serve dim sum dishes are carted around the restaurant for customers to choose their orders while seated at their tables. Eating dim sum at a restaurant is usually known in Cantonese as going to "drink tea" as dim sum is typically served with tea. Similar small dishes exist in other varieties of Chinese cuisine, sometimes known in English as dianxin, from the pronunciation of the same two characters for "dim sum" in Mandarin.

Dim sum is usually linked with the older tradition from yum cha , which has its roots in travelers on the ancient Silk Road needing a place to rest. Thus, teahouses were established along the roadside. An imperial physician in the third century wrote that combining tea with food would lead to excessive weight gain. People later discovered that tea can aid in digestion, so teahouse owners began adding various snacks.

The unique culinary art dim sum originated with the Cantonese in southern China, who over the centuries transformed yum cha from a relaxing respite to a loud and happy dining experience. In Hong Kong, and in most cities and towns in Guangdong province, many restaurants start serving dim sum as early as five in the morning. It is a tradition for the elderly to gather to eat dim sum after morning exercises. For many in southern China, yum cha is treated as a weekend family day. More traditional dim sum restaurants typically serve dim sum until mid-afternoon. However, in modern society it has become common place for restaurants to serve dim sum at dinner time, various dim sum items are even sold as take-out for students and office workers on the go.

CUISINE

A traditional dim sum brunch includes various types of steamed buns such as cha siu baau, dumplings and rice noodle rolls , which contain a range of ingredients, including beef, chicken, pork, prawns and vegetarian options. Many dim sum restaurants also offer plates of steamed green vegetables, roasted meats, congee porridge and other soups. Dessert dim sum is also available and many places offer the customary egg tart. Dim sum can be cooked by steaming and frying, among other methods. The serving sizes are usually small and normally served as three or four pieces in one dish. It is customary to order family style, sharing dishes among all members of the dining party. Because of the small portions, people can try a wide variety of food.

TEA SERVICE

The drinking of tea is as important to dim sum as the food. The type of tea to serve on the table is typically one of the first things the server asks dining customers. Several types of tea are served during dim sum :

Chrysanthemum tea – Chrysanthemum tea does not actually contain any tea leaves. Instead it is a flower-based tisane made from chrysanthemum flowers of the species Chrysanthemum morifolium or Chrysanthemum indicum, which are most popular in East Asia. To prepare the tea, chrysanthemum flowers are steeped in hot water (usually 90 to 95 °C after cooling from a boil) in either a teapot, cup, or glass. However, Chrysanthemum flowers are often paired with Pu-erh tea, and this is often referred to as guk pou or guk bou .

Green tea – Freshly picked leaves only go through heating and drying processes, but do not undergo fermentation. This enables the leaves to keep their original green color and retain most natural substances like polyphenols and chlorophyll contained within the leaves. This kind of tea is produced all over China and is the most popular category of tea. Representative varieties include Dragon Well and Biluochun from Zhejiang and Jiangsu Provinces respectively.

Oolong tea – The tea leaves are partially fermented, imparting to them the characteristics of both green and black teas. Its taste is more similar to green tea than black tea, but has a less “grassy” flavor than green tea. The three major oolong-tea producing areas are on the southeast coast of China e.g. Fujian, Guangdong and Taiwan.

Bo-lei tea or Pu-erh tea – The tea has undergone years of fermentation, giving them a unique earthy flavor. This variety of tea is usually compressed into different shapes like bricks, discs and bowls.

Scented teas – There can be various mixtures of flowers with green tea, black tea or oolong tea. Flowers used include jasmine, gardenia, magnolia, grapefruit flower, sweet-scented osmanthus and rose. There are strict rules about the proportion of flowers to tea. Jasmine tea is the most popular type of scented tea, and is often the most popular type of tea served at yum cha establishments.

The above teas are produced in most of China. Chinese tea bushes (Camellia sinensis) are cultivated in the mountain areas of tropical and subtropical regions or wherever there is a proper climate, sufficient humidity, adequate sunshine and fertile soil. Chinese tea is classified in many ways, e.g., quality, method of preparation or place of production. The main processing methods include fermentation (oxidation), heating, drying and addition of other ingredients like flowers, herbs or fruits. These help to develop the special flavor of the raw tea leaves.

GUNFIRE

Gunfire is a British cocktail made of black tea and rum. It has its origins in the British Army and is also used as a name for early morning tea in the army.

BRITISH ARMY

It is unknown when Gunfire was concocted; however it is known that it was mixed by British Army soldiers during the 1890s. Gunfire is served by Officers and Non-Commissioned Officers to lower ranks before a morning attack (as a form of Dutch courage) and as a celebration before a Passing out parade. It is also traditionally served to soldiers in their beds by their officers on Christmas Day at Reveille if they are deployed over Christmas. Individual regiments may carry out the ritual on other days; for example, within the Royal Tank Regiment Gunfire is served on Cambrai Day. During the Korean War, members of the American Military Police Corps were given Gunfire by British soldiers under the guise of it being normal tea after a recovery mission. This led to intoxication of the MPs who then drove an Armoured recovery vehicle and some army jeeps into a camp gate as a result of Gunfire consumption.

AUSTRALIAN AND NEW ZEALANDER ARMIES

In Australia and New Zealand on ANZAC Day, a version of Gunfire with black coffee instead of tea is served to soldiers after dawn services as part of the "gunfire breakfast".

CIVILIANS

Gunfire has also been made and drunk outside of military circles. Gunfire was served to participants of British reality programme, Bad Lads Army by the Non-Commissioned Officers before their passing out parade, mirroring the same procedure in the British Army. Gunfire is also drunk by Australian civilians as well to commemorate ANZAC Day. A similar drink, particularly in the German-speaking world is Jagertee.

RECIPE

Gunfire consists of one cup of black tea with one shot of rum, which is then stirred in the cup.

YUM CHA

Yum cha also known as ban ming is a southern Chinese style morning or afternoon tea, which involves drinking Chinese tea and eating dim sum small dishes . There are various ways in which the food can be served. In highly patronized restaurants, staff will commonly wheel around heated trolleys serving small yum cha dishes that have been prepared in advance in the kitchen; this manner allows customers receive hot, fresh items quickly. In less busy places, ordering dim sum will typically involve marking an order sheet with the desired quantity of items to be ordered; this practice eliminates unnecessary preparation of dishes which may otherwise not be ordered.

Yum cha in Cantonese Chinese literally means "drink tea", while ban ming is a more poetic term meaning "tasting of tea". In the US and UK, the phrase dim sum is often used in place of yum cha; in Cantonese, dim sum refers to the wide range of small dishes, whereas yum cha, or "drinking tea", refers to the entire meal. Cantonese-speaking communities in Canada, however, still tend to use the phrase "yum cha".

PREVALENCE

The Cantonese Chinese term yam cha or yum cha primarily refers to the tradition of morning tea in Cantonese cuisine exemplified by the traditional tea houses of Guangzhou . Due to the prevalence of Cantonese cuisine outside China, the Cantonese yum cha tradition can be found in many parts of the world. By analogy, yum cha is also used to refer to morning or afternoon teas in other Chinese cultural traditions, even though such meals have different native names.

DIM SUM

Similarly to a Western morning or afternoon tea, despite the name, yum cha is focused as much on the food items served with the tea as the tea itself. These food items are collectively known as "dim sum", a varied range of small dishes which may constitute or replace breakfast, brunch or afternoon tea. Dishes are usually steamed or fried and may be savoury or sweet. They include steamed buns such as char siu baau, assorted dumplings, siu mai, and rice noodle rolls, which contain a range of ingredients, including beef, chicken, pork, prawns and vegetarian options. Typical desserts include egg tarts, sai mai lo and mango pudding. Many yum cha restaurants also offer plates of steamed green vegetables, roasted meats, congee porridge, and soups.

Dim sum can be cooked by steaming and frying, among other methods. The dim sums are usually small and normally served as three or four pieces in one dish. It is customary to share dishes among all diners on the same table. Because of the small portions people can try a wide variety of food.

Traditionally, the cost of the meal was calculated based on the number and size of dishes left on the patron's table at the end. In modern yum cha restaurants, dim sum servers sometimes mark orders by stamping a card on the table. Servers in some restaurants even use different stamps so that sales statistics for each server can be recorded.

EATING UTENSILS

Methods of tea-tasting have undergone considerable changes since the Tang Dynasty and the use of covered teabowls is one development of note. The simple yet practical design of covered teabowls makes them an excellent utensil for tea-tasting. The fine painting on these bowls also enhances their aesthetic value. Nowadays a small teacup is used instead in most dim sum restaurants.

Chopsticks, a pair of small even-length tapered sticks, are the traditional eating utensils used for yum cha. Chopsticks are commonly made of plastic in the Chinese restaurants. Held between the thumb and fingers of the working hand they are used as tongs to pick up portions of food or to sweep rice and small particles of food into the mouth from the bowl. Dim sum dishes are served in small portions and therefore convenient for eating with chopsticks. Many rules of etiquette govern the proper use of the chopsticks.

Toothpicks are frequently used between courses, as it is believed that the aftertaste of one course should not be allowed to ruin one's enjoyment of the next course. Toothpicks are ideal and socially acceptable for picking up those meal items which often defy the best chopstick approach, such as slippery button mushrooms and jellyfish slices served with sesame oil.

CUSTOMS AND ETIQUETTE

It is customary to pour tea for others before filling one's own tea cup. It is most gracious to be the first to pour tea.

When tea drinkers tap the table with two fingers of the same hand, an action known as 'finger kowtow', it means thanks. According to a just-so story, this gesture recreates a tale of imperial obeisance and can be traced to the Qianlong Emperor of the Qing dynasty, who used to travel incognito. While visiting the Jiangnan region, he once went into a teahouse with his companions. In order to maintain his anonymity, he took his turn at pouring tea. His stunned companions wanted to kowtow for the great honour but to do so would have revealed the identity of the emperor. Finally, one of them tapped three fingers on the table (one finger representing their bowed head and the other two representing their prostrate arms) and the clever emperor understood what he meant. From then on, this has been the practice.

YIXING CLAY

Yixing clay is a type of clay from the region near the city of Yixing in Jiangsu Province, China. Its use dates back to the Song dynasty when Yixing clay was first mined around China's Lake Taihu. From the 17th century on, Yixing wares were commonly exported to Europe. The finished stoneware, which is used for teaware and other small items, is usually red or brown in color. Also known as zisha ware, they are typically unglazed and use clays that are very cohesive and can form coils, slabs and most commonly slip casts. These clays can also be formed by throwing. The best known wares made from Yixing clay are Yixing clay teapots.

"Yixing clay" is often used as an umbrella term to describe several distinct types of clay used to make stoneware:

- Zi sha or zi ni: this stoneware has a purple-red-brown color.
- Zhu sha or zhu ni: reddish brown stoneware with a very high iron content. The name only refers to the sometimes bright red hue of cinnabar. There are currently 10 mines still producing zhu ni . However, due to the increasing demand for Yixing stoneware, zhu ni is now produced in very limited quantities. Zhu ni clay is not to be confused with hong ni , another red clay.
- Duan ni: stoneware that was formulated using various stones and minerals in addition to zi ni or zhu ni clay. This results in various textures and colours, ranging from beige, blue and green to black.

MINERAL COMPOSITION

Zisha is a mixture of kaolin, quartz and mica, with a high content of iron oxide. It is mined principally at Huanglongshan and Zhaozhuangshan and has a somewhat sandy texture. The process of preparing the clay is lengthy and was traditionally regarded as a trade secret. Typical firing temperature is between 1100C - 1200C in an oxidizing atmosphere.

MANUFACTURING

The raw materials for yixing clay are buried deep underground, sometimes under heavy sedimentary rock formations. When excavated, it is usually located within stratified layers of other clays. The seam of yixing zisha can be as thick as a several decimeters, up to a meter. Yixing clays consist of fine iron-containing silt, with mica, kaolinite and varying quantities of quartz and iron ores as its main mineral constituents. Processing of raw zisha yixing clay involves removing the clay from the underlying strata, drying it under the sun in open stalls, and then pulverizing the dried clay pieces into fine particles. The clay powder then undergoes air screening to isolate clay particles of the finest grit size. The screened clay is then mixed with water in a cement mixer to a thick paste, piled into heaps, and vacuum processed to remove air bubbles, in addition to some moisture from the clay mixture. The quality and quantity of water in yixing clay is critical in that it determines the quality of the stoneware products produced. After this processing, the resulting clay is then ready to be used.

The appearance of yixing products, such as its colour or texture, can be enriched and altered through the addition of various metal oxides into the yixing clay, through the manipulation of firing temperatures, and also from regulating the kiln atmosphere .

USE

Yixing teawares are prized because their unglazed surfaces absorb traces of the beverage, creating a more complex flavour. For these reasons, yixing teawares should never be washed using detergents, but rather with water only, and connoisseurs recommend using each tea vessel for one kind of tea or sometimes even one variety of tea only.

Early pots were designed for travel use hence you will see the simple classical look of the pots produced during the Ming dynasty. Most tea drinking enthusiast will have one teapot for travel use, these tend to be less expensive and compact in design. It was not until during the mid-Qing dynasty that tea connoisseurs started to use the pot at home and the artisan begin to form them into different shape and sizes. Many exotic forms were conceived. Vessels were decorated with poetic inscriptions, calligraphy, paintings and seals were incised onto the surface of the teapots.

WAGASHI

Wagashi is a traditional Japanese confectionery which is often served with tea, especially the types made of mochi, anko (azuki bean paste), and fruits. Wagashi is typically made from plant ingredients. In Japan the word for sweets, okashi, originally referred to fruits and nuts. China learned from India how to produce sugar and began trading it to Japan. The trade increased and sugar became a common seasoning by the end of the Muromachi period. Influenced by the introduction of tea and China's confectionery and dim sum, the creation of wagashi took off during the Edo period in Japan.

TYPES

- Anmitsu: chilled gelatinous cubes (kanten) with fruit
- Amanattô: simmered azuki beans or other beans with sugar, and dried - amanattô and nattô are not related, although the names are similar.
- Botamochi: a sweet rice ball wrapped with anko (or an, thick azuki bean paste)
- Daifuku: general term for mochi stuffed with anko
- Dango: a small, sticky, sweet mochi, commonly skewered on a stick
- Dorayaki: a round, flat sweet consisting of castella wrapped around anko
- Hanabiramochi: a flat, red and white, sweet mochi wrapped around anko and a strip of candied gobo

- Ikinari dango: a steamed bun with a chunk of sweet potato and anko in the center, it is a local confectionery in Kumamoto.
- Imagawayaki : anko surrounded in a disc of fried dough covering
- Kusa mochi: "grass" mochi, a sweet mochi infused with Japanese mugwort , surrounding a center of anko
- Kuzumochi
- Kuri kinton: a sweetened mixture of boiled and mashed chestnuts
- Manjû: steamed cakes of an surrounded by a flour mixture, available in many shapes such as peaches, rabbits, and matsutake mushrooms
- Mochi: a rice cake made of glutinous rice
- Monaka: a center of anko sandwiched between two delicate and crispy sweet rice crackers
- Oshiruko : a hot dessert made from anko in a liquid, soup form, with small mochi floating in it
- Rakugan: a small, very solid and sweet cake which is made of rice flour and mizuame
- Sakuramochi: a rice cake filled with anko and wrapped in a pickled cherry leaf
- Taiyaki: like a kaitenyaki, a core of anko surrounded by a fried dough covering, but shaped like a fish
- Uirô: a steamed cake made of rice flour and sugar, similar to mochi
- Warabimochi: traditionally made from warabi and served with kinako and kuromitsu
- Yatsuhashi: thin sheets of gyûhi , available in different flavors, like cinnamon, and occasionally folded in a triangle around a ball of red anko
- Yôkan: one of the oldest wagashi, a solid block of anko, hardened with agar and additional sugar
- Akumaki: one of the confections of Kagoshima Prefecture

CLASSIFICATION

Wagashi are classified according to the production method and moisture content. Moisture content is very important, since it affects shelf life.

- Namagashi —contains 30% or more moisture
  - Jô namagashi is a very soft and delicate, seasonally varying namagashi, in various, often elaborate, shapes and colors, often reflecting seasonal plants. Some stores will have many dozens over the course of a year.
  - Mochi mono
  - Mushi mono
  - Yaki mono
  - Hiranabe mono

  - Ôbun mono
  - Nagashi mono
  - Neri mono
- Age mono Han namagashi —contains 10%–30% moisture
  - An mono
  - Oka mono
  - Yaki mono
  - Hiranabe mono
  - Ôbun mono
  - Nagashi mono
- Neri mono Higashi —contains 10% or less moisture
  - Uchi mono
  - Oshi mono
  - Kake mono
  - Yaki mono
  - Ame mono

TURKISH TEA

Turkish tea is a type of tea that is popular mainly throughout Turkey and the Turkish diaspora, as well as in Northern Cyprus and some Balkan countries.

Turkish tea, called çay , is black tea which is consumed without milk, is produced on the eastern Black Sea coast, which has a mild climate with high precipitation and fertile soil. Turkish tea is typically prepared using two stacked kettles called "çaydanlýk" specially designed for tea preparation. Water is brought to a boil in the larger lower kettle and then some of the water is used to fill the smaller kettle on top and steep several spoons of loose tea leaves, producing a very strong tea. When served, the remaining water is used to dilute the tea on an individual basis, giving each consumer the choice between strong (Turkish: koyu; literally "dark", tavþan kaný — a deep brownish red or weak (Turkish: açýk; literally "light"). Tea is drunk from small glasses to enjoy it hot in addition to showing its colour, with cubes of beet sugar.

In 2004 Turkey produced 205,500 tonnes of tea (6.4% of the world's total tea production), which made it one of the largest tea markets in the world, with 120,000 tons being consumed in Turkey, and the rest being exported. Furthermore, in 2004, Turkey had the highest per capita tea consumption in the world, at 2.5 kg per person—followed by the United Kingdom . Tea is grown mostly in Rize Province on the Black Sea coast.

BACKGROUND

Tea is an important part of Turkish culture, and is the most commonly consumed hot drink, despite the country's long history of coffee consumption. Offering tea to guests is part of Turkish hospitality. Tea is most often consumed in households, shops and mostly kýraathane, which is social congregation of Turkish men. Despite its popularity, tea only became the widely consumed beverage of choice in Turkey starting in the 20th century. It was initially encouraged as an alternative to coffee, which had become expensive and at times unavailable in the aftermath of World War I. Upon the loss of southeastern territories after the fall of the Ottoman Empire, coffee became an expensive import. At the urging of the founder of the republic, Atatürk, Turkish people turned more to tea as it was easily sustainable by domestic sources. Turkish tea is traditionally offered in small tulip-shaped glasses which are usually held by the rim, in order to save the drinker's fingertips from being burned, as the tea is served boiling hot.

TURKISH HERBAL TEAS

In Turkey, herbal teas are not considered true teas, but are used as herbal medication. They are mostly popular with tourists with apple , rose hip , and linden flower being the most popular flavors. Sage tea (ada çayý, literally meaning "island tea") is most popular in the Mediterranean coastal region. In Turkey, herbal teas for most ailments can be found in local herbal shops. Dried herbal leaves, petals, shoots, etc. are sold in loose-leaf according to each customer's need and taste.

TRACING TEA

Tracing Tea is a documentary about the history and culture of tea production and consumption in Asia and Europe. Produced by Maximum Exposure Productions, the focus of the series will be on a driving expedition from Darjeeling, the spiritual home of tea in India, back to the UK, the drink's greatest importer. The journey, which covers some 10,000 miles , will use two modified Indian auto rickshaws and aims to achieve two world records: a) the longest journey ever completed in a tuk-tuk and b) the world's highest tea party - on top of Pakistan's Khunjerap pass .

The official route comprises seventeen different countries as follows: India, Pakistan, China , Kyrgyzstan, Kazakhstan, Uzbekistan, Turkmenistan, Iran, Turkey, Bulgaria, Serbia, Hungary, Austria, Germany, the Netherlands, France and finally Britain. The filming begins in July 2008 and will run until February the following year.

PROGRAMME CONTENT

In each country, the series will look at how tea is produced and consumed, and how over the centuries, particularly in traditionally Islamic societies, the consumption of tea has become not just a drink but an elaborate social institution. Tea producing countries to be covered: India, Pakistan, Iran, Turkey and even the UK . Concurrently, the documentary will present themes and sights on a wider theme, from Islamic Architecture in Iran and Turkish cuisine to nomadic sports in Central Asia. Central to all the above will be the 'reality' element, following the progress of the three presenters as they battle terrain, the elements and the mechanical implications of doing the journey in auto-rickshaws.

TEAM

The core crew comprises the director, Max Lovell-Hoare, producers Robin Forestier-Walker and Sophie Ibbotson and two main presenters Andrew Daynes and Sam Datta-Paulin. Previous crew members included Urvi Nagrani, Rosie Dell, and Aijan Temiralieva. Andrew Daynes has already made a five-month reconnoitre of the route in reverse, with a team of three, using a London taxi.

THAI TEA

Thai tea, also known as Thai iced tea is a Thai drink made from strongly brewed Ceylon tea. However, due to Ceylon tea's high price, a locally grown landrace version of Assam known as Bai Miang with added food coloring is commonly used. Other ingredients may include added orange blossom water, star anise, crushed tamarind seed or red and yellow food coloring, and sometimes other spices as well. This tea is sweetened with sugar and condensed milk and served chilled.

Evaporated milk, coconut milk or whole milk is generally poured over the tea and ice before serving to add taste and creamy appearance. However, in Thailand, condensed milk and sugar are mixed with the tea before it is poured over ice and then topped with evaporated milk. In Thai restaurants worldwide, it is served in a tall glass, though when sold from street and market stalls in Thailand it is more typically poured over the crushed ice in a clear plastic bag or tall plastic cups. At markets, it can be seen to be mixed through pouring the tea at heights of about 4 feet back and forth. It can also be made into a frappé at more Westernised vendors.

It is popular in Southeast Asia and in many American restaurants that serve Thai food. Although Thai tea is not the same as bubble tea, a Southeast and East Asian beverage that contains large tapioca pearls, Thai tea with tapioca pearls is a popular flavor of bubble tea.

VARIATIONS

Cold:

- Dark Thai iced tea Thai tea served chilled with no milk content, sweetened with sugar only. The concept is based on traditional Indian tea which is used as main ingredient.
- Lime Thai tea Similar to Dark Thai iced tea, but flavored with lime as well as sweetened with sugar. Mint may also be added.

Hot:

- Usually, Thai people drink Thai hot tea in the morning, frequently with Yiu Ja Guoy or pathongko as it is called by most Thais.
- Thai hot tea Thai tea served hot.
- Dark Thai hot tea Thai tea served hot with no milk content, sweetened with sugar only.

TERERÉ

Tereré (of Guaraní origin) is an infusion of yerba mate / erva-mate , similar to mate but prepared with cold water rather than with hot, and in a slightly larger vessel. It is originally from Paraguay, and is found also in northeastern Argentina and southern and western Brazil. When hot (mate), the Guaraní people call this infusion ka'ay, where ka'a means herb and y means water. The scientific name of yerba mate is Ilex paraguariensis.

The vast majority of people in Paraguay take their tereré with water-infused herbs such as mint "menta-i" or lemongrass. When not prepared with plain cold water, citrus fruit juices are usually used, although this practice varies depending on the region. While mixing fruit juices with tereré is common in northeastern Argentina, Brazil and Paraguay, Lime and orange juices are used in Argentina and Paraguay; lime and pineapple juice are more common in Brazil. Tereré taken with juice is commonly called "tereré ruso" because this practice is more common with Slavic immigrants in the southeast of Paraguay and northeast of Argentina, than with Spanish- and Guaraní-speaking Paraguayan people.

In some parts of Argentina, tereré is seen as a lesser form of mate, and its drinking hardly ever follows the traditional drinking mate ritual. In Paraguay, by contrast, tereré is considered preferable to mate on a hot day.

First invented by the Guaraní natives who lived in Paraguay and western Brazil (Mato Grosso do Sul) territory that used to belong to Paraguay before the war of the Triple Alliance, Tereré was spread by the dwellers of that region, and for centuries was a social beverage. People usually prepare one jar of natural water and a "guampa" / "cuia" with a "bombilla" / "bomba" which is shared among the group of people. Since Paraguay and Mato Grosso do Sul have a very hot climate, this drink is excellent to refresh the body and can be considered a very low-calorie, non-alcoholic beverage. Additionally, it is an important ritual signifying trust and communion.

Guampas/Cuias are gourds that can be made from animal horns, usually made from ox horns. The new cuia or the one not used for some time needs to be watered before use, because the lower part of the cuia is usually capped with a round piece of leather. By filling the cuia with water the leather expands, covering all leaks in the cuia. Many cuias are capped using a wooden cork as well. Cuias are also made from wood and covered in leather or steel. In central

Argentina, it is common practice to drink terere from a small metal cup, as these receptacles don't absorb and retain fruit juice and are easier to clean.

Bombillas/Bombas are metal straws with a filter at the end. Erva-mate/ Yerba is placed at the bottom of the cuia, and as water is added the bomba drinks from the cuia the clear green liquid.

Tereré is a typical drink from Paraguay. It is very popular among young and old people. It is common to see students going to school with a guampa and a vacuum flask under their arms, filled with ice and water, as summer temperatures can reach 40 to 45 °C . It is used as a popular remedy to treat ailments such as headache, stomachache and diarrhea by adding medicinal herbs into the water.

TERERÉ PREPARATION

Most preparations of Tereré begin by filling a cured yerba mate gourd, guampa, or cup 1/3 to 1/2 full of yerba mate. Then, ice cubes, fruit, sweeteners, and/or herbs are added. The drinking vessel is then filled with juice or ice water. At this point, a mate straw is inserted through the mixture to the bottom of the vessel, while a thumb is kept over the bombilla tip in order to keep it from becoming clogged. The liquid is consumed through the straw, and refilled as desired.

Some of the more popular juices and fruits to use in Tereré are lime, orange, lemon, mango, peach, or grapefruit. Some of the more popular herbs to use in Tereré are mint, lemongrass, or lemon verbena. Of course, any herb or juice could be added, even beer and lime cola have their supporters.

TEH TARIK

Teh tarik is a hot milk tea beverage which can be commonly found in restaurants, outdoor stalls and kopi tiams within the Southeast Asian countries of Malaysia and Singapore. Its name is derived from the pouring process of 'pulling' the drink during preparation. It is made from black tea, condensed milk and/or evaporated milk. It is also considered as the national drink of Malaysia.

The mixture is poured back and forth repeatedly between two vessels from a height, giving it a thick frothy top. This process cools the tea to optimal drinking temperatures, and helps to thoroughly mix the tea with the condensed milk. It is also done to give the tea a better flavour. This is often compared to the decantering of toddy to improve its flavour. Locally and regionally sourced tea used for teh tarik are not of the highest grade. Despite the strong aroma from the common Ceylonese variety, the taste is rather acrid and generally would not go well with a little cream or fresh milk like other fine tea. Hence condensed milk or evaporated milk is used to ensure any overpowering taste is well balanced by the creamy flavour of the thickened milk.

CULTURE

Since colonial times, Teh tarik has been a popular Malaysian Indian cuisine for several people in Malaya and Singapore. The origins of teh tarik can be traced to Indian-Muslim immigrants in the Malay Peninsula who set up drink stalls at the entrance of rubber plantations after World War II to serve the workers there. Traditionally, teh tarik has been seen served with the Roti Canai which has become a popular breakfast set by Malaysians until today.

An element of showmanship exists in the preparation of teh tarik. The ability to drag a long stream of tea above the heads of the patrons without giving them a shower is an amusing novelty for the locals and tourists alike.

In Malaysia, there are occasions where teh tarik brewers gather for competitions and performances to show their skills. Teh tarik has become recognised along with nasi lemak as part of the food and beverage heritage of Malaysia by the Malaysian government ministry.

VARIATIONS

- Teh-C refers to tea made with unsweetened evaporated milk, unlike the traditional teh tarik which is made with sweetened condensed milk. Vendors however will add sugar to Teh-C unless specifically asked not to. The Malay word for nil or zero, kosong, is used to indicate that the drinker does not want their beverage to be sweetened.
- Teh-O refers to the same tea used but without any addition of dairy entirely. Like Teh-C, sugar is usually added except when Teh-O kosong is specified where once again no sugar is added.
- Teh halia, ginger tea which is also found in various coffee shops.
- Teh Tarik kurang manis, can be translated to "less sweet Teh Tarik"
- Chiam is a combination of tea and coffee.

TEA-PICKING OPERA

A Tea-picking opera is a form of musical entertainment. It originally derived from the tea-growing region around Mount Jiulong in the South of Jiangxi Province in China, where the tea pickers would sing lengthy songs to each other whilst undertaking the monotonous task of tea-picking. The songs were often in three parts with different groups of pickers, singing different parts. These folk songs were gradually adopted by performing troupes of singers and eventually became known as Tea picking operas. In Taiwan, the Hakka people have a form called the "Hakka Tea-picking Opera".

TEA PET

Tea Pet, also known as a tea lover's pet, refers to a type of traditional craft from China. Its history can be traced back to the Yuan dynasty . It is usually made of a special clay, "Zisha" , from the region near Yixing in Jiangsu province, China. Just like Yixing teapots made of clay, tea pets are unglazed, so that they are mostly monochromatic with a rough surface. As the name implies, a tea pet is like a pet raised by tea lovers. Tea lovers in China raise a tea pet by placing it on the tea tray during tea time and pouring out the tea over it. The most popular figure of the tea pet is the "pee-pee boy", which is used to judge whether the water is hot enough to make tea. Tea pets are also molded into zodiac animals or Chinese mythical creatures such as dragons, Pixiu, Qilin, etc., to symbolize good luck, fortune and happiness, as well as historical or mythical characters such as Guanyin, Maitreya and Zhu Geliang.

The birthplace of Tea pets, Yixing, was first famous as the birthplace of "Yixing clay" in Song dynasty . With the popularity of the Yixing clay teapot, Yixing became a production center for teaware since then. As early as the Yuan dynasty , the tea pet was made in Yixing for the first time. Teapot artisans started to mold Yixing clay into various mythical creatures or animals as mascots for tea lovers. However, probably because the ancient Chinese scholars believe in a saying that "excessive attention to trivia saps the will", there is very limited literature on the development of tea pets in Chinese history. Nowadays, even with the development of the technology— it is not hard to find alternatives to Yixing clay as the material to make tea pets— the production is still concentrated in Yixing region.

TYPES

A Tea pet is handmade using Zisha and unglazed, so it is usually in the natural color of Yixing clay. There are three kinds of Yixing clay: purple clay, red clay, and green clay. A tea pet can be made of either of one of these Yixing clays, or a mixture of two clays to produce different colors.

Purple clay, is the principle clay, which turns red-brown or dark-brown after firing. The artisan often adds iron clay to achieve a better color effect.

Red clay, also known as "Stock Yellow", has a much higher contraction percentage than purple clay, and shows a vermillion color after firing. Because of its high contraction percentage, red clay is more suitable for making small-sized objects like tea pets and teapots.

Green clay has a color similar to duck egg shell, and appears off-white after firing. It is rarer than purple clay, and more expensive on the market.

RAISE A TEA PET

The way that tea lovers raise a tea pet is quite similar to the way that they maintain a Yixing clay teapot. It is a very time-consuming process. They often pour leftover tea over their tea pets, and use a tea brush to wipe the surface of the tea pet to help it absorb the tea evenly. Tea pets should be only rinsed with water, without using soap or any dishwashing liquid, to ensure a progressive increase in the fragrance and color of the tea. After prolonged maintenance, over months or years, the tea pet will absorb the tea, and its appearance will become glossier along with the aroma of the tea.

PEE PEE BOY

Pee pee boy is the most representative and popular figure among tea pets. It is about 3 inch tall with a red-brown color. The ability to judge the water temperature is its most useful quality contributing to the popularity of the form. To do this, a tea lover immerses it in cold water until it is filled up and shakes it to make sure it is filled at least 50 percent with water; after that, hot water poured over the pee pee boy will make it "pee" if the water is hot enough. The hotter the water, the further it pees. The principle behind the pee pee boy is thermal expansion of air. It is designed to be hollow with only one tiny opening, so that water can flow into pee pee boy at a slow pace and will not drip out until sufficient hot water is poured on its head. When hot water is poured over its head, the air in the pee pee boy expands,resulting the water to be squeezed out through the tiny opening. With this feature, the pee pee boy soon became the most popular figure in tea pets. And recently, more and more artisans also utilize this technology on other tea pets, such as water-breathing dragons or a water-spraying longevity tortoises, to provide more choices for tea lovers.

TEA PARTY

A tea party is a formal, ritualized gathering for the small meal called afternoon tea. Formal tea parties are often characterized by the use of prestige utensils, such as porcelain, bone china or silver. The table is made to look its prettiest, with cloth napkins and matching cups and plates. In addition to tea, larger parties may provide punch, or in cold weather, hot chocolate. The tea is accompanied by a variety of foods that are easy to manage while in a sitting room: thin sandwiches, such as cucumber or tomato, cake slices, buns or rolls, cookies, biscuits and scones are all common.

FORMAL TEAS

The afternoon tea party was a feature of great houses in the Victorian and Edwardian ages in the United Kingdom and the Gilded Age in the United States, as well as in all continental Europe (France, Germany, and above all in the Russian Empire). The formal tea party still survives as a special event, as in the debutante teas of some affluent American communities.

In the older version, servants stayed outside the room until needed. Writing in 1922, Emily Post asserted that servants were never to enter the room unless rung for, to bring in fresh water and dishes or to remove used dishes. This was partly due to the rigidity of social convention at the time, but it also reflected the intimate nature of the afternoon tea. Proving the truth of 18th-century author

Henry Fielding's quip that "love and scandal are the best sweeteners of tea", the custom of banning servants from the drawing room during tea shows the hostess's desire to encourage free conversation among her guests. Most of the formalities of that age have disappeared, particularly since World War II, when economic changes made household servants a rarity, but afternoon tea can still provide a good opportunity for intimate conversation and a refreshing light meal.

"KETTLE DRUMS"

A less formal large afternoon party for tea was known during the 18th and 19th centuries as a "kettle drum". A widespread but possibly false folk etymology suggests that the name "kettle drum" may have originated in the informal tea gatherings hosted by British camp officers' wives during East India Company rule or the British occupation of India, during which kettle drums are claimed to have served as tea tables in the camps. Alternately, "kettle drum" may have been an amalgam of "drum" — 18th-century slang for a vivacious party — and "kettle" for the tea served. At kettle drums, guests traditionally came for short periods and left at will, mingled and conversed with little formality, and partook of tea, chocolate, lemonade, cakes, and sandwiches. Guests were expected to dress for ordinary daytime visiting, but not more formally.

TEA PARTY OF ALICE

In the chapter "A Mad Tea-Party" in Alice's Adventures in Wonderland, Alice becomes a guest at a tea party along with the March Hare, the Hatter, and a sleeping Dormouse who remains asleep for most of the chapter. The other characters give Alice many riddles and stories, including the famous 'Why is a raven like a writing desk?'. The Hatter reveals that they have tea all day because time has punished him by eternally standing still at 6 pm . Alice becomes insulted and tired of being bombarded with riddles and she leaves claiming that it was the stupidest tea party that she had ever been to.

TEA LADY

A tea lady is a woman in an office or working environment, whose sole job is to provide beverages (chiefly tea) and light snacks during the allocated tea break. Tea ladies are a mainly British custom. They entered the mainstream in the UK during the Second World War, when tea ladies were used in an experiment to boost efficiency in workplaces for the war effort. They had such a hugely positive effect on morale they became commonplace in all areas of work, mobile canteens even serving military units on exercises. They were present in a works canteen or perhaps came round with a trolley, on which was usually a tea urn full of either hot tea or hot water, and perhaps a selection of cakes and buns.

In the past they were often used as stereotypical figures in British comedy, with a tea lady usually portrayed as a harassed, overweight, middle aged woman in a uniform and cap, or as a very pretty recipient of all sorts of lewd comments from the workforce, as in the Carry on film Carry on at your Convenience.

The character Mrs Doyle in the 1990s sitcom Father Ted, set in Ireland, had the role of a tea lady, and became known for relentlessly forcing the other characters to drink the tea she'd made for them, possibly a reference to the fact that the Irish are the biggest tea drinkers in the world averaging 4 to 6 cups per person per day. Tea ladies in general were a frequent target of illusory "cuts" and "economies" in Yes Minister, frequently conjured up by Nigel Hawthorne's character Sir Humphrey Appleby. However, tea ladies were never once seen onscreen for the whole five-series run of the show.

In association football, media and fans often blame "the tea lady" for leaking information about player transfers to the public, regardless of whether or not team managers actually have a tea lady.

DEMISE

This profession began to die out in the late 1970s to early 1980s when tea ladies began to be replaced by private catering firms and vending machines, as businesses expanded and women moved into higher paying jobs. The tradition of the tea break, from which the role of tea lady rose, has itself declined, also offering a possible explanation why tea ladies are not commonly found today.

A survey carried out by the group National Opinion Polls in 2005 showed that of those workers who drink more than four cups of tea a day, only 2% of them receive it from a tea lady, whereas 66% received it from an urn, and 15% from a vending machine.

TEA HOUSE

A tea house or tea room is an establishment which primarily serves tea and other refreshments. Although its function varies widely depending on the culture, tea houses often serve as centers of social interaction. Some cultures have a variety of distinct tea-centered houses or parlors that all qualify under the English language term "tea house" or "tea room."

ASIA

In China and Nepal, a tea house is traditionally a place which offers tea to its consumers. People gather at tea houses to chat, socialize, and enjoy tea, and young people often meet at tea houses for dates. The Guangdong style tea house is particularly famous outside of China especially in Nepal's himalayas. These tea houses, called chálou serve dim sum and these small plates of food are enjoyed alongside tea. In Japanese tradition a tea house ordinarily refers to a private structure designed for holding Japanese tea ceremonies. This structure and specifically the room in it where the tea ceremony takes place is called chashitsu. The architectural space called chashitsu was created for aesthetic and intellectual fulfillment.

In Japan during the Edo period, the term "tea house" could also refer to a place of entertainment with geisha or as a place where couples seeking privacy could go. In this case the establishment was referred to as an ochaya, which literally meant "tea house". However, these establishments only served tea incidentally, and were instead dedicated to geisha entertainment or to providing discreet rooms for visitors. This usage is now archaic. Contemporary Japanese go to modern tearooms called kissaten on main streets to drink black or green tea as well as coffee.

In Central Asia the term tea house could refer to Shayhana in Kazakh, Chaykhana in Kyrgyz and Choyxona in Uzbek, which literally means a tea room. In Tajikistan. The largest tea houses are Orient Tea house or Chinese Tea house, Orom Tea house in (Isfara) town. On the 15th anniversary of Independence in Tajikistan, the people of Isfara town presented Isfara Tea house to Kulyab city for its 2700th anniversary on September 2006. Tea houses are present in other parts of Central Asia, notably in Iran and also Turkey. Such tea houses may be referred to, in Persian, as Chay-Khaneh, or in Turkish, çayhane - literally, the "house of tea." These tea houses usually serve several beverages in addition to tea.

In Arabic-speaking countries such as Egypt, establishments that serve tea, coffee and herbal teas like karkade are referred to as ahwa or maqha and are more commonly translated into English as coffeehouse.

EUROPE

Tea drinking is a pastime closely associated with the English. Tea first arrived in England during Cromwell's protectorate and soon became the national drink, with tea drinking a national pastime for the English. As early as 1784, La Rochefoucauld noted that "[t]hroughout the whole of England the drinking of tea is general". Nevertheless, Anna Russell, Duchess of Bedford, is credited with the invention of afternoon tea. By 1840, it had spread to other parts of English society with the female manager of London's Aerated Bread Company attributed with innovating the first commercial public tearoom.

Thomas Twining opened the first known tea room in 1706, which still remains at 216 Strand, London. In 1787, the company created its logo, still in use today, which is thought to be the world's oldest commercial logo that has been in continuous use since its inception. Under Associated British Foods since 1964, Stephen Twining now represents the company's tenth generation. In 2006, Twinings celebrated its 300th anniversary with a special tea and associated tea caddies. Twining's is a Royal Warrant holder . There is a long tradition of tea rooms within London hotels, for example, at Brown's Hotel at 33 Albemarle Street, which has been serving tea in its tea room for over 170 years. In the UK today, a tea room is a small room or restaurant where beverages and light meals are served, often having a sedate or subdued atmosphere. A customer might expect to receive cream tea or Devonshire tea, often served from a china set, and a scone with jam and clotted cream – alternatively a High tea may be served. In Scotland teas are usually served with a variety of scones, pancakes, crumpets and other cakes. In a related usage, a tea room may be a room set aside in a workplace for workers to relax and take refreshment during work-breaks. Traditionally a staff member serving food and beverages in such a tea room would have been called a tea lady. Tea rooms are popular in Commonwealth countries, particularly Canada, with its harsh winters, when afternoon tea is popular. The menu will generally have similar foods to in the UK, but with the addition sometimes of butter tarts or other small desserts like nanaimo bars or pets de sœurs. Tea is commonly consumed in other Commonwealth countries alone or in the British fashion.

In France, a tea room is called Salon de thé, and pastries and cakes are also served. It seems having a separate tea house was a culture in many countries in Europe. In Germany, one Teehaus was particularly famous during the Third Reich era where the German Dictator Adolf Hitler used to have his daily walk and tea on Mooslahnerkopf hill near his residence Berghof, in the Bavarian Alps. Hitler's tea house was a cylindrical structure built in the woods. In the Czech Republic, the tea room culture has been spreading since the Velvet Revolution 1989 and today, there are nearly 400 tea rooms in the country , which is according to some sources the largest concentration of tea rooms per capita in Europe. In Eastern Europe, countries like Latvia are located at the crossroads of trade routes between Western and Eastern Europe, and tea came both from the East and West. One example of mixed tea is a new type of tea room - Club tea culture. For example - a tea club Goija.

RELATIONSHIP TO 19TH CENTURY TEMPERANCE MOVEMENT

The popularity of the tea room rose as an alternative to the pub in the UK and US during the temperance movement in the 1830s. The form developed in the late 19th century, as Catherine Cranston opened the first of what became a chain of Miss Cranston's Tea Rooms in Glasgow, Scotland, and similar establishments became popular throughout Scotland. In the 1880s, fine hotels in both the United States and England began to offer tea service in tea rooms and tea courts, and by 1910 they had begun to host afternoon tea dances as dance crazes swept both the U.S. and the UK. Tea rooms of all kinds were widespread in Britain by the 1950s, but in the following decades cafés became more fashionable, and tea rooms became less common.

4

Tea Processing

Tea processing is an agro based non conventional industrial sector with activities concentrated mostly in hilly areas. The environmental status of the sector has not been compiled so far even though it is already recognized as an industrial activity by the Government. The Central Pollution Control Board has therefore undertaken a project for preparation of "Comprehensive Industry Document on Tea Processing Industry". The report covers the national and international scenario of tea processing industry, and its categorization, process details, environment aspects related to wastewater, air pollution and solid waste management along with the proposed environmental standards.

TEA PROCESSING INDUSTRY – NATIONAL AND INTERNATIONAL SCENARIO

NATIONAL SCENARIO

One of the important agro-industrial crop, which supports Indian economy largely, is tea. Tea is indigenous to India and is an area where the country can obviously take a lot of pride. Not only because tea is a major player in the earning of foreign exchange but also contribute substantially to the country's GNP. The technology of tea cultivation and processing has been developed within the industry aided by applied research which was largely funded by the tea companies themselves. This has helped to grow tea industry as a technically competent entity and also allowed the generation of more academic and fundamental investigations which might bring future rewards. Now, when global market has opened, all commodities are facing stiff competitions tea is not an exception. In fact, one of the major reason for the refusal of Indian tea by the consumer world wide is the cost of production and marketing.

SCOPE AND METHODOLOGY OF THE PROJECT

SCOPE OF THE PROJECT

The scope of work includes the followings:

- Inventorization of the large, medium and small tea processing industries, their capacity & location.

- Classification of the type of tea processing units based on different technologies & processes capacities.
- Description of process, pollution control technologies, guidelines, and environmental standards followed in other countries, and
- Crop patterns of tea cultivation – and its environmental impact on soil, ground water, nearby surface water bodies etc.

METHODOLOGY

Representative industries were selected from large, medium and small tea manufacturing industries based on different technologies, processes & capacity for in-depth studies. The aspects covered in the detailed studies are as follows:

- General description of manufacturing process with detailed flow diagram;
- Genesis of pollution & identification of pollutants in effluents, emissions and solid waste and their probable quantities
- Preparation of overall material, energy & water balance;
- Identification of sources of fugitive emissions;
- Identification of appropriate technologies for the control of air pollution and (iii) fugitive emissions;
- Guidelines for identification of waste disposal site;
- Identification of suitable clean technologies for tea manufacturing units under Indian conditions;
- Formulation of standards for wastewater discharge and air emissions in terms of concentration and quantum of pollutants on the basis of the findings and suggesting of the rationale for achieving the formulated standards.

TEA PLANTATION AND RELATED ASPECTS

SOIL

Tea is grown in a wide range of soil types found in tropical, subtropical and temperate climate conditions. These soil types have developed from diverse parent rock material and under different climate conditions. In China, Indonesia, Sri Lanka, South India, Turkey and Georgia (USSR) tea is mostly grown on sedimentary soils derived from gneiss or granite. In north-east India, except in Darjeeling, tea is grown on flat alluvial lands which occupy the vast area of the Brahmaputra Valley in Assam. Peat soils that have been drained are successfully used in Cachar to grow tea. But in Kyoto and Kanaya, the main areas of Japanese tea, the crop is grown on soil types derived from volcanic ash and in Taiwan, tea occupies a tract of tertiary rocks derived from a residual formation. However, all these soils have one common characteristic i.e. they grows best in acid soils. Despite the diversity of soil types on which tea is grown, all the soils exist in high rainfall conditions, as this is the most important climatic factor for successful tea growing. These soils get formed under special type of weather combined with intensive leaching of the products of weathering. The degree of leaching and hence the character of the resulting soil depends not only on the rainfall but also on the temperature. Because of differences in temperature, soils formed in tropical climates are likely to have certain characters different from those formed in the sub-tropical or semi-tropical conditions. The tea areas in tropical climates experience minimal temperature variations as compared with those in subtropical climates.

RAINFALL

The relationship between tea production and rainfall is well known. Tea grows well in areas having a precipitation of about 1,150 to 8,000 mm. The effect of rainfall is perhaps more manifested by its influence on moisture status of the soil and in inducing vegetative growth. Therefore, distribution of rainfall is as important as the total annual rainfall. It is difficult to say what exactly is the water requirement of tea in different phases of its growth and development. It is natural to expect that water requirement of tea would vary according to the prevailing environmental conditions, but it is reasonably assumed that tea on an average may transpire 900 mm per annum. Ideally, the crop water requirement should be such that it does not cause any stress in the plant system.

HUMIDITY

Humidity is of importance in tea physiology primarily because of its influence in determining the loss of moisture by evapo-transpiration. High humidity reduces water loss, but low humidity increases it, and also increases water stress during rainless drought period. In relation to temperature, low humidity may be advantageous because by increasing transpiration rate it also reduces leaf temperature, though the effect will be more pronounced in unshaded conditions. In north-east India the humidity level generally remains high during the harvesting period and this is generally considered to be conducive to growth.

AIR TEMPERATURE

Tea is grown in tropical to temperate conditions, under a regime of air temperature that varies between –8°C and 35°C; the suitable temperature for growth being the one common to the habitat of the plant, that is, where it grows. But photosynthetic rate of tea is at maximum between 30°C and 35°C, falls rapidly at 37°C, and between 39°C and 42°C there is virtually no net photosynthesis. There is also no uptake of carbon dioxide at about 42°C; respiration may continue up to about 48°C but the leaf is irrevocably damaged.

Therefore, there has to be an optimum range of temperature for growth and productivity of tea. In north-east India the air temperature remains above 30°C for long periods during harvesting season, and hence the need for shade to keep the temperature below a level where it cannot affect photosynthesis adversely. It is important that a high vapour-pressure deficiency is also associated with high temperatures; and it would be useful to partition the effects of these two components on plant growth.

LAND PREPARATION

Environmental considerations discourage the replacement of rain forest by tea. However, it is worth recalling that a properly managed tea plantation causes only a temporary disturbance of many of the characteristics of the rain forest. Whilst tea cannot be rain forest as the plant species and associated fauna are different, the stream flows and evapo-transpiration revert to the same as the forest when the tea is mature. Clearance of any type of vegetation disturbs the soil and leaves it unprotected. Clearance and planting must be timed with the minimum of intervening delay. The mechanics of planting depend on the density of vegetation to be removed. Trees should be killed in advance of clearing to minimise the risk of root disease remaining alive to infect the tea. Trees should be ring barked or frilled; the trees need inspection at intervals because the bark tends to regrow across such cuts. Trees may take as long as three years to die. Chemicals used at one time to kill trees have been withdrawn for safety reasons. Trees and other heavy material should be moved off the field. The accumulation of quantities of organic material in heaps must be avoided as it may raise the pH of the soil to an unacceptable level. Light material may be ploughed in, or burnt if well spread out. Large fires must be avoided as the ash will raise the pH of the soil. A low pH will be more tolerant of ash and vegetation than soil with pH close to the upper limit for tea.

EROSION CONTROL AND DRAINS

Drains should be put in as needed on flat land. These are particularly necessary on swampy ground and should be deep enough to allow for the settlement which often occurs when such land is drained. Roadway formations should be leveled with drains and culverts put in place.

WATER REQUIREMENT

Theoretically, the water requirement of tea is the amount needed to balance the water lost by transpiration, which is a continuous process. Though rainfall is essential for growth & productivity of tea, it must be of the right quantum and intensity. The effect of rainfall is perhaps more manifested by its influence on moisture status of the soil. Normally a part of rainwater is lost by surface run off, percolation and evaporation, only a portion of the heavy & high intensity rain percolates to the root zone of tea to cause appreciate physiological effects. Soil conditions also react with rainfall in more than one way. The ability of the soil to withstand long dry season without adverse effect depends on the quantity of rain water that the soil is capable of retaining and can make available to the plants. This depends, apart from the capacity of the soil to retain water, on the spread and depth of tea roots. On the other hand, excessive rainfall in shallow soil and depressions may cause localised waterlogging - the latter affects both root efficiency and photosynthesis. Heavy rainfall may also affect pollination. However, the adverse effects of either excess rainfall or lack of adequate rainfall can be compensated by drainage and irrigation respectively. Normally in tea plantation water is supplied in the form of spray by water sprinkler. This measure controls the soil moisture and also allows coverage of large sectors from limited resources. Drainage and irrigation are very closely linked with each other. A well-drained land will provide better conditions for tea roots to grow deeper and to proliferate well. A deep rooting system will thus be able to explore a much bigger volume of soil for water and nutrients and will be able to withstand drought better and longer. On the other hand, in poorly drained lands the roots remain very shallow and tea starts showing symptoms of soil moisture stress much earlier. This also demands for more frequent irrigation for survival. To obtain optimum benefits of irrigation, it is therefore essential to provide adequate drainage in the area to be irrigated.

SOIL CHEMISTRY

Acidity

It is generally agreed that optimum range of soil pH for the tea plant is 5.0 to 5.6. Soil those above pH 5.6 require correcting with acid materials such as Aluminium sulphate, Sulphate of ammonia, elemental Sulphur. On the lower end of the scale tea grows in as low a pH as 4.0 or below. The main causes of soil acidity are Hydrogen and Aluminium ions. In an intensively leached soil in which tea grows best, cations are replaced by H+ from the water molecule and can be washed off the mineral. This is the main source of acidity in tea soils. Aluminium ions also react with water, produce acidity. Nitrogenous fertilizers and decomposed organic matter in soils also supply Hydrogen ions cause acidity to soil. In case of soil pH less than 5.0, use of lime to bring the pH to about 5.0 is quite convenient.

Organic Matter

Different Organic Matters accumulate into the soil from plant & animal tissue through well decomposition, acts as the binding material joining soil particles to form a stable soil structure. Under natural condition it is the main source of plant nutrient. Tea soils vary in their OM contents ranging from less than 1% to over 30%. However, the tea plant once established and well maintained generates a lot of OM from leaf fall and prunings (10 - 20 tonnes biomas produced per hectare per year).

WATER LOGGING

Water logging is a physical factor, which can render a soil to be unsuitable for tea. This is very important because it is known that tea roots will not grow when the soil or subsoil is water logged for a prolonged period. Thus, a soil which has poor drainage should be avoided.

Water Holding Capacity

The water-holding capacity of a given tea soil and its availability to plants is dependent on soil textural composition, their aggregation and the rooting depth of the tea plant. The coarser the texture (higher content of coarse material) the less the water-holding capacity. The less the water-holding capacity the shorter times such a soil will be able to supply water to the tea plant, especially in an extended period without rain or irrigation.

FERTILIZER

Fertilisation is an important part of the normal intensive production of tea. Tea is normally grown as a long-term monoculture. Without applied fertilizer the supply of nutrients available in the soil will become exhausted leading to mineral deficiencies in the plants, severe reduction in yield and ultimately, to the death of plants and a degraded plantation. Nutrients are removed from a field of tea in a number of ways. There is an irreversible loss in the crop. The net loss of nutrients in old leaves and prunings depend on the extent to which these items are retained in the field. Soil erosion, drainage of excess water containing nutrients in solution and decomposition to gases create further losses. Uncontrolled weeds also absorb substantial quantities of nutrients. In addition to oxygen, hydrogen and carbon which are obtainable in air & water, tea plant requires fourteen (14) mineral nutrients for its ideal growth. These are Nitrogen , Potassium , Magnesium , Manganese , Copper (Cu), Aluminium , Boron and Chlorine (Cl). Of these essential nutrients mentioned above, the first nine are considered principal nutrients as they are required in relatively larger amounts. The details of plant nutrients is being presented below in Fig.

Functions Of Crop Nutrients

Primary nutrients :

- Nitrogen (N)
  - Induces vegetative growth
  - Essential constituent of protein, chlorophyll, hormones, alkaloids and Vitamins
  - Governs the ultisation of Phosphorus and Potassium

PLANT NUTRIENTS

MAJOR NUTRIENTS		MINOR NUTRIENTS	
Primary Nutrients	**Secondary Nutrients**	**Micro Nutrients**	**Beneficial elements**
Carbon	Magnesium	Zinc	Aluminium
Hydrogen	Calcium	Manganese	Silicon
Oxygen	Sulphur	Boron	
Nitrogen		Iorn	
Phosphorus		Copper	
Potassium		Molybdenum	
		Chlorine	

- Phosphorus (P) :
  - Stimulates root growth
  - Carrier of energy
  - Essential for synthesis of starch, protein and fat
  - Constituent of nucleic acids and enzymes
  - Related to cell division and development
- Potassium (K) :
  - Essential for frame development
  - Regulates water metabolism in plants
  - Provides resistance against drought, frost, pests and diseases
  - Acts as an accelerator of enzyme action
  - Counteracts the ill-effects of excess Nitrogen

Secondary nutrients :

- Calcium (Ca) :
  - Constituent of cell wall
  - Necessary for normal mitosis
  - Essential to active growing points
  - Activator of enzymes
- Magnesium (Mg) :
  - Constituent of chlorophyll
  - Imparts dark green colour to leaves
  - Helps in synthesis of carbohydrates, protein, fat and vitamins
- Sulphur (S) :
  - Constituent of Sulphur bearing amino acids
  - Helps in stabilizing protein structure

Deficiency Symptoms Of Nutrients

Primary nutrients :

- Nitrogen (N) :
  - Young flush shows a lighter than normal green colour-yellowing
  - Marked decline in crop
- Phosphorus (P) :

- Mature leaves loose their glossiness and appear dull
- Stunted appearance of plants
- Potassium (K) :
  - Marginal necrosis in mature leaves
  - Premature leaf fall leaving a crown of tender foliage
  - Development of thin and twiggy wood
  - Poor recovery after plucking and pruning

Secondary nutrients:

- Magnesium :
  - Seen on mature leaves
  - Leaves appear bright yellow with a conspicuous inverted dark green "V" down the
  - Mid rib
- Sulphur :
  - Chlorotic symptoms occur in younger and recently formed leaves – Tea
  - Yellows
  - Lateral extension of leaves is restricted leading to formation of narrow laves
- Micronutrients :
- Boron :
  - Terminal bud remains dormant and die early
  - Clusters of small shoots develop in the upper axils
  - Development of translucent oil spots on the lower surface of mature leaves
  - Formation of cork mound on the leaf stalk

Organic Fertilizer

Organic manure is applied to tea as green manure or as one of many other organic materials. Organic materials limits the deterioration of the soil and improves soil fertility and structure. A bed of humus is formed which supports an active micro fauna which speeds decomposition converting nutrients to an available from. Erosion is reduced and the infiltration of water improved.

Green Manure

The tea plants provide some green manure from leaf-fall. The leaves of shade also form green manure. Leguminous or other trees planted in lines as windbreaks are trimmed frequently in some areas; leguminous trees, planted as hedges separated by a few lines of young tea as temporary shelter, can provide up to 100 kg Nitrogen per hectare per annum within 25-30 tonnes organic matter. These close hedges have to be removed before the tea comes into production. Removal of the prunings takes a large quantity of nutrients out of the plant soil system. This loss will not be replaced by normal fertilizer applications. If prunings are removed regularly the tea will develop nutrient deficiencies, yield will fall and the health of the plants will decline. Grice and Malenga reported that removal of prunings produced a 4% reduction in crop yield in the cycle after a down prune.

TEA PROCESSING - GREEN PLANT TO FINISHED PRODUCT

In India, black is manufactured and used. Green leaf tea is also manufactured which is mainly exported. This document is mainly eccentric to black tea manufacturing.

NURSERY

A good plant can be developed and grown only under a regime of good agromanagement in the nursery, till it is transplanted to the field. Therefore, a nursery must not only to provide the most optimum conditions for growth of the plants but the chosen sites should also be convenient for the area to be planted to minimise any damage to the plants during transit, and indeed minimise transport cost. It should be reasonably leveled with no risk of waterlogging during heaviest of rains. The soil should be chemically suitable with a good structure. A deep top soil normally ensures a good depth of fertile soil in the bed. A useful nursery requires all weather access, shade, drained beds and water. Sandy loam soil is the best for raising a nursery on orthodox beds though virgin soil will provide the best medium. Extreme soils at best are avoided. On heavy soils seedling growth is rather slow, though by adding sand or cattle manure, plant growth in these soils could be hastened within limits.

Dark coloured top soil is normally rich in fertility but soil very poor in organic matter may do with a little dry cow dung or a fine compost blended in. A useful practice is to sieve the soil through No. 3 mesh to remove undesirable elements like old bits of roots. It is equally important that the soil is not drying out, turning muddy or is compressed as these might affect the growth of the plants. To facilitate growth, 500 gm of super-phosphate per cubic metre soil would be advantageous; other fertilisers may be incorporated if analysis shows a specific need. In any case, no soil should be used which has a high proportion of eelworms. Current trend is to grow tube nurseries rather than bed nurseries as the management of plants is much easier when they are grown in tubes. Although size of the tube is a matter of convenience, an optimum size should ensure good growth. Six to eight weeks before their use the sleeves are filled ideally with good textured medium-loam soil which ensures that water does not accumulate within the sleeves. If more than half of the soil is fine silt the sleeve soil will have to have sand. The pH value of the soil should remain between 4.5 and 5.5. Extreme values outside this range should be adjusted with lime or aluminium sulphate. For nursery plants, a low level of shade is essential but they should give a clearance of 25 cm above sleeve tops. Shade used in Assam and the Dooars consists of 1.30 x 1.30 m bamboo basket laid across the bed fences. Winter planting of nurseries can be started without shade, but a light mulch is needed to prevent drying out of the bed. Eventually, a shade has to be provided not only to prevent desiccation of the soil but also to provide appropriate growth stimuli to the nursery plants as well. But more than shade per se, a proper regulation of shade is essential for the growth of the nursery plants.

In general, as the nursery plants get themselves established shade, can be selectively reduced, and is finally removed when it interferes with the growth sequence of the plants. The latter is done when the nursery plants have produced four to five good leaves but prior to that the process has to be gradual to allow the growing nursery plants in getting accustomed and adjusted to intense sunlight. Shade is removed differentially first when there is a two leaf growth. This is done first on a cloudy day and finally the shade is removed when the plants have four to five leaves growth. At higher elevations, 1,400 m and above, a polythene tent supported on bamboo strips and hoops 45 m above the bed would be useful. Application of fertilisers will increase the rate of growth. When soil in the tube or bed is fertile, nitrogen alone will have a significant effect.

Phosphate is also effective is most situations. A soluble NPK mixture for foliar application (10 kg urea plus 20 km ammonium phosphate plus 8 kg murate of potash) is applied from unfolding of two leaves till plants have eight leaves. Weeds can be a problem when plants are small. At the very incipient stage weeds can be removed physically, though a some herbicide will effectively control the weeds for a longer duration. Applications of insecticides, acaricides and fungicides will depend on specific pest and disease situations. Disbudding or tipping is useful for encouraging laterals, but this should be done cautiously to avoid any damage to growing plants. This is best done when plants are 30 to 35 cm high with step up.

FIELD PLANTING

From all accounts planting is a most crucial operation as it basically determines the development and productive level of tea throughout its economic life. Wrong planting of good planting materials is doubly unproductive as investment is lost both on account of producing the plants and in the failure to put them up for productivity. Therefore, care, planning and refinement of techniques are essential for long-term benefit. The very fundamental to planting, apart from good planting materials, is that the planting hole should be of adequate size. The size of the hole partly depends on the consistency of the soil but a diameter of 45 cm and a depth of 75 cm is ideal; the hole could be dug in three phases - the first phase is dug to 15 cm involving cutting of top soil with a hoe and placing it separately; the second phase involves digging up to 30 cm which involves subsoil layer -the excavated portion is also maintained separately. In the third phase the balance 30 x 15 cm across is loosened with posthole auger or deep forking. The size of hole also depends on the consistency of the soil but in general a standard size of 45 x 75 cm ought to be maintained.

When the holes are being dug the crumb structure of the soil is often damaged, but alternate wetting and drying of the puddle system corrects this. However, if the excavated holes are exposed for a long time without refilling, the excavated soil may be lost due to erosion. Therefore, the holes should be dug with sufficient time in hand to allow the soil to settle down. For closer spaced planting and heavy soils, an alternative to holing is trench planting. Trenches 30 cm wide and 45 cm deep are dug along rows either manually or mechanically. Trenches provide a lower volume of soil for root expansion. The holes are refilled two or three weeks before planting with top soils mixed with dry old cattle manure (4 kg) or 5 kg well-composted organic matter mixed with 30 to 40 gm of superphosphate. Since the initial growth of the plants is dependent on their food reserve rather than mineral nutrition of the soil, organic manures would be valuable. However, care must be taken to avoid having a layer of compost at the bottom of the pits as it may cause localised water logging and ensure that high concentration of mineralising organic matter does not produce a high level of pH.

BRINGING UP OF YOUNG TEA PLANTS AND BUSH FORMATION

When a young plant is established in the field it is essential to train or modify it to form a low bush of good spread and of convenient height for plucking. The essence therefore is to establish a good frame and ensure a good coverage of ground within the shortest possible time. This is achievable either by de-centering and pruning, or by bending or pegging, or by a combination of de-centering and bending. The decentering and pruning combination involves cutting off of the main stem at a height between 7 cm and 20 cm - the few laterals remaining on the stem are cut at 35 cm above ground level. This excision stimulates the plants to throw out more laterals from the lower portion of the stem, thus increasing the number of pruning sticks. The plants are next pruned within two years after the first prune for de-centering, and this is done about 5 cm above the last prune mark. This part of the operation, that is, decentering, can be done as soon as the plants start producing new growth after transplanting, and in more vigorous plants this can be done even earlier than in weaker plants.

In hilly areas, where plant growth rate is generally slow de-centering is done in spring or autumn after planting. A drawback of de-centering process is that it might cause high mortality unless the full vigour of the plants is assessed. The argument in favour of bending and pegging is that they do not cause heavy mortality during the operation, provide quick ground cover, apart from producing a substantial yield by the second year. For vigorous plants, pegging can be done two to three weeks after planting. Shoots that are 30 to 60 cm high are bent at angles 30° to 45° to the ground, and are then pegged down to the ground. All lateral shoots produced subsequent to the bending are also bent and pegged. But bending could be difficult with overgrown plants and is risky in un-shaded areas because of scorching risks. Moreover, in contour planting, it is always desirable to have the 'bends' along the contour, preferably turned towards west or south. The pegged main steam and the laterals are first pruned 30 to 40 cm from the ground within two to three years, and are then raised by 5 cm in each successive prune till the permanent frame height is reached.

Three variants of pegging are practiced, at least in north-east India. First, bending with single pegs, which is a laborious process involving repeated pegging using single pegs either of bamboo, wood or bamboo. It is always difficult to keep the pegs in position as often they are removed by the growing pressure of the branches. In the ring pegging method after the initial bending of the plant by a single peg, or following a de-center, the laterals that emerge are allowed to spread out by putting a ring of bamboo strips 25 to 30 cm diameter in the middle. This ring is often kept in position by three to four pegs but the process is rather cumbersome. In 'rail' pegging method, the laterals and re-growth are repeatedly bent pushing them beneath two rails of bamboo strips. Though easy to operate, the main drawbacks here are the difficulties in maintaining the desired angle. Yet a third approach towards frame formation is to de-centre first and then go for pegging. This essentially involves de-centering of the main stem with the laterals being bent to the sides and pegged. Subsequently, pruning is done at 35 to 50 cm from the ground and then the frame is raised by 5 cm at each successive prune till a permanent frame is established.

One advantage of this system is that it is easier to train the branches as less number of pegs are needed. On account of decentering in the first year there could be an initial yield loss of about 18 percent, but the subsequent rise in crop well compensates the earlier losses. As pointed out earlier, the main objective in rearing young tea is to divert apical dominance of the main stem and force the plants to develop well-spread lateral branches. To achieve this twain objective a "low tipping method" was advocated. This technique involves total avoidance of pruning in the first four years of planting, centering out within 10 weeks after planting and initial plucking at a low level of 40 to 50 cm from the ground. The first pruning is considered only when the plants are above 85 cm high at the end of the fourth or fifth year with pruning height varying between 55 and 65 cm from the ground. This technique resulted in high yield in the early years but it failed to give a food frame so essential for long-term productivity of tea.

Therefore, by giving a prune, rather than totally avoiding it in the early years, a good frame could be established. With the introduction of bending or pegging, good ground cover, ideal frame and early crop are possible to achieve. Whether pegging would be ideal with high plant populations, irrespective of the growth habits of different cultivars, is not clear. Three other systems used in varying degrees are centering out, lung-pruning and de-budding. Centering out involves cutting out of the main stem at a height of 15 to 20 cm to force lateral growth from below the cut.

This operation can be done six to eight weeks after planting or when the plants have produced a few flushes or leaves indicating that they have established themselves. Since the operation is rather drastic, it is most important that starch reserve of the plants is at its optimum. If, per chance, only a few leaves have developed, then it is better to avoid this system and go for either lung-pruning or de-budding.

Lung-pruning or thumb-pruning as it is popularly known, is a mechanical operation. When plants are between 15 and 20 cm high, they are held between the thumb, and the stem is half broken, making sure that one side is kept intact for movement of food and nutrient. The upper part of the stem is then bent towards the ground and the broken portion is removed only after the plant produces a flush of new growth. Unlike centering out operation, lung-pruning can be done even if starch content is not at its optimum.

De-budding, as the name implies, consists of removal of the apical bud and all buds in leaf axils, 15 to 25 cm above the ground. This ensures that no laterals grow from the upper part of the plant and after laterals have grown from the lower portion, the stem is removed at 15 to 20 cm. De-budding is an extremely delicate operation and can even be done in the nursery in situations where growth of laterals is a problem.

It is difficult to generalise on the relative merits and demerits of different systems of raising young tea because the principal source of variation, that is, the growth characteristics of different cultivars have not been considered in most studies, despite marginal evidence that some tea varieties behave alike under a particular system of bush management. This needs thorough elaboration using cultivars of different genotypic origin. Nonetheless, by adopting one or other system, the growth rate of young tea can be enhanced ensuring early yield.

On relative efficacies of the four principal methods — low tipping and step up plucking; low tipping with frame forming prune and step up plucking; pegging and step up plucking; and pegging after de-centre with step up plucking, the general views are:

- Low tipping produces good early crop but frame formation is poor, and frame formation prune 18 months after planting produce good frame with high yields from fifth year onwards.
- Pegging has an advantage in terms of crop increase only in the two initial years. Thereafter, this advantage is lost and the system's overall

performance equals the combination of low tipping and frame formation pruning.

- De-centering before pegging causes an immediate crop loss but this is wellcompensated later. This apart, pegging by itself may cause serious sun scorch which in some cases may even remain permanently on the frame.

NECESSITY OF PRUNING

Perhaps from a productive point of view no field operation is as important as pruning is, as it basically helps in maintaining the plant as low bush in a phase of continuous vegetative growth. Pruning therefore is considered as "at best a necessary evil" as it both stimulates and controls growth. The other principal objectives of pruning are :

- Shaping the tree to make the best use of space between trees.
- Reestablishing the initial vigour of the shoot system to stimulate growth, and control undue rise in bush frame and plucking table, that is, control growth.
- Removing dead, diseased and overage wood, thus help rejuvenate bushes that have crossed the period of maximum productivity.
- Providing an environment within the plant which is conductive to maximum crop production but minimises the spread of pests and diseases.
- Having the crop formed where it can be harvested easily and cheaply.
- Renewing the actively growing branches and maintaining sufficient volume of maintenance foliage to meet physiological needs of the plant.
- Maintaining quality in made tea.

It is generally accepted that bushes are best pruned during December to January and must not be rested earlier than October. Resting during a phase of slow growth is essential to create a favourable carbohydrate balance, but this principle does not hold good for all countries, that is, the timing of pruning varies depending on local growing conditions. The correct timing for pruning is decided basically from an assessment of the growth rate of the plants. It should be at a point when growth has sufficiently slowed down to create a favourable carbohydrate balance, so that, a layer of foliage could grow and get established well before the hot spell. The other factors that influence pruning time are prevention from potential dangers of sun-scorch, drought or attacks from pests and diseases, apart from the more mundane considerations like the re-growth time desired and the crop planned. In summarising, pruning is a means for reducing the number of stalks on the bushes, so that, an optimum quality of wood is left for giving the best and most vigorous shoots the following year. The amount of wood left will again vary according to the growth characteristics and vigour of the cultivars but leaving 3 to 4 cm of wood may be useful under certain situations.

CUTTING TYPES

Pruning

Collar Pruning

Collar pruning is the severest form of pruning, requiring as it does, cutting the bole of the bush near or at the ground level. In some cases, the cut may even be given slightly below the ground level to ensure protection of the exposed tissues against die-back. Recovery from this severe form of pruning is greatly influenced by carbohydrate reserve. By and large, collar pruning is suitable for tea that is characterised by a well marked dormancy period. The main objective is to provide a new fame to the plant by removing old and unproductive branches affected by diseases and pests, particularly in situations where uprooting of unproductive bushes is not possible. Tea recovering from collar pruning also needs special care and management. In heavy pruning, which is a variant of collar pruning, the bole is cut at 15 to 45 cm above the ground for frame renewal. But this kind of hard pruning can only be done after serious consideration of plant health and carbohydrate reserve. A debilitated plant exposed to extreme forms of pruning may cause further depletion of food reserve without producing any compensatory maintenance foliage, and in the process the frame formation would suffer.

Medium Pruning

The main objective of medium pruning is to reduce height of the bushes, so that, they do not exceed 80 cm. Continuous picking of shoots on the top of a bush gradually increases the height of the bushes, apart from causing congestion on the top with weak and twiggy branches. This drain in the physiological resources which otherwise could have been utilised for increasing productivity. Therefore, to bring back the plants to a productive level, stimulate new wood and renew maintenance foliage, medium pruning becomes a useful means. Though relatively less severe than collar or heavy pruning, medium pruning is generally not done below 50 cm, and certainly not lower than what is necessary to remove the knots and congested top. As with collar pruning, a major determinant for medium pruning is to decide that the plants are ready to take this drastic treatment, that is, whether they have adequate carbohydrate reserves, and whether chances of scorching are minimal. In north-east India, this operation is ideally carried out during end December to end January, but before this operation is carried out, plants must be rested for six to eight weeks.

Top or Light Pruning

This is the standard method of recurring pruning in Assam and the Dooars. The main objective of this pruning is to renew the leaf bearing branches, apart from cleaning out the bush. Normally top pruning is done 2 to 3 cm above the last prune; in effect this system of pruning leaves a considerable amount of foliage on the plants though much of it defoliates naturally; but under certain conditions of growing this foliage may even remain infact and carry on normal physiological functions. The technique of light pruning therefore essentially involves a uniform cut-across followed by a cleaning out operation to remove thin unproductive branches and dormant shoots. Sometimes only a cut-across, that is, removal of top part of light pruned tea is practiced to invigorate growth without undue removal of foliage.

Lung Pruning

The essence of lung pruning is to prune selectively, so that, all leaves below the level plucking are left and the peripheral branches are not cut. These left out branches have turned lungs though in no way are they connected with respiration per se. This system of pruning ensures that sufficient amount of actively synthesising foliage is left on the bush so that a carbohydrate balance is maintained to enable optimal growth of the plants. The use of this system of pruning is rather restricted, and leaving a few 'breathers' at the time of medium pruning have generally been beneficial for stimulating growth.

Skiffing

Skiffing is the lightest form of pruning involving as it does removal of certain amount of growth above the previous pruning level. It essentially is the levelling of the foliage either by giving a cut in the green stem in the lighter forms of skiffing, or, into the red wood in harder forms of skiffing. The object of skiffing is not only to have an early start of crop production but also to achieve more of the high quality first flush and total crop as well. Skiffing helps in thickening the pruning wood and improving the general health of the bushes.

Tipping

Immediately after pruning, the plants enter into a phase of unrestricted growth and if this unrestricted growth is allowed, most of the shoots would mature as branches. To prevent this, the new growth has to be controlled, so that a uniform plucking table is formed. Tipping therefore is an operation aimed at forming a level and a flat plucking surface, and filling it by quick production of secondary branches.

PLUCKING GREEN TEA

Plucking or picking denotes harvesting of tea crop. Plucking of young shoots, the two leaves and buds, is an complex art consuming about 15 percent of the total cost of production of tea and much of the quantity and quality of harvested crop depends on the standard of plucking. It involves removal of young growing shoots comprising the apical bud, the internodes and two or three leaves below it, which together constitute the crop. The tender shoots removed are the sinks as they cannot support their normal growth. They are harvested at regular intervals to stimulate successions of a new crop of shoots. The growing shoots in turn are dependent on the assimilates from the mature foliage where they are produced, that is, the source. Consequently, an ideal plucking system must take cognizance of the intricate relationship between the sink and the source. In practical terms, the amount of maintenance foliage to be retained is of prime consideration in determining the system of plucking to be adopted as it is the maintenance foliage that provides the assimilates for the production of shoots. The growing bud exerts maximum pulling force for metabolites from the maintenance foliage; the relative sink capacity of a shoot of one leaf and a bud and two leaves and a bud being 75 and 35 percent, respectively. Therefore, indiscriminate removal of large buds and one leaf and a bud would adversely affect the health of bushes. The initial growth of leaves following pruning occurs at the expense of carbohydrates formed during the previous growth and stored in roots and older wood. These new shoots develop from the dormant buds on the stubs that are left on the plant during pruning. As explained earlier, these crops of new shoots are the primaries and the two covers on the dormant buds are known as the scale leaves or cataphylls or janams. The emerging leaves are arranged in spirals around the stem in such a manner that the first five leaves form a series which encircle the stem twice, while the sixth leaf appears above the first of the previous series.

PLUCKING STANDARD

The standard of plucking is unequivocally important as more than anything else, it determines the quality potential of the processed leaf. It is generally accepted that fine plucking makes the best quality tea as this form of plucking consists picking of only tow leaves and a bud and plucking of these flushes is regarded as fine plucking. In a broader sense, fineness of plucking may also involve harvesting all one and a bud, two and a bud and single banjhis over the janam, apart from the fact that fine plucking also calls for plucking at short intervals usually at seven days intervals. If still shorter rounds of plucking are practiced to get leaves with attributes for high quality, then it may cause severe depletion of 'sinks' on plucking surface as the shoot will not have opportunities for optimal growth so necessary before plucking.

Consequently, over a period of time total yield may decline but quality of tea will certainly improve. At the other extreme of fine plucking are the medium, coarse or very coarse plucking. These three systems of plucking greatly increase the quantity of harvested crop, but only at the cost of quality. Medium plucking, for example, involves harvesting of large shoots with two leaves and a bud, three leaves and a bud plus the single banjhis or dormant leaves. Coarse plucking involves removal of all shoots with two leaves and a bud, three leaves and a bud, four leaves and a bud and double banjhis. Very coarse plucking removes all shoots with three leaves and a bud and even shoots with four leaves and a bud. Coarse plucking can be done at 8 to 10 or even 12 days intervals, and because of the increase in the size of shoots, the weights of harvest also increase proportionately with the length of the shoots harvested.

Indeed, a linear relationship, exists between the size of shoots with different combinations of leaves and a bud and their weights. In terms of 'sink' capacity of pluckable shoots, shoots with one leaf and a bud have the highest sink capacity - therefore, their indiscriminate removal along with larger buds may adversely affect the physiological balance causing a decline in the production of subsequent flushes. From this point of view at least standard plucking would appear to be ideal.

PLUCKING INTERVAL

For efficient harvesting of shoots plucking interval is most crucial. The interval is ideally determined by the rate of a leaf unfolding. In north-east India, the general practise is to harvest shoots at 7 days intervals on the assumption that by this time the levels of various biochemical constituents necessary to make tea of quality would reach their optima. Any increase or decrease in plucking interval could significantly influence the dynamic metabolic system, even leading to alterations in leaf physiology. Therefore, plucking rounds at shorter intervals with fine plucking can reasonably assure a better quality of made tea than teas plucked at longer intervals. Thus, 5 days plucking rounds can give good quality of made tea because of a lower content of ash and fibre, though black teas made from 7 days pluckings would have a balance of VFC , ash, soluble solids and caffeine and therefore would have an optimum quality. Teas manufactured from 9 to 11 days pluckings are generally of low quality possibly because of a decrease in soluble solids, with a significant increase in crude fibre.

It is equally important that theaflavin also vary with the maturity of leaves at 5 to 11 days plucking intervals. Apart from quality, plucking interval also affects overall quantum of harvestable shoots. Generally, as the plucking interval or length is increased, there is a corresponding increase in yield of both light pruned and unpruned tea, but quality deteriorates. This happens because longer rounds increase the quantum of coarser leaf and banjhi leaves, and both have a negative effect on quality. This drawback notwithstanding, plucking interval is often extended to 8 to 10 days in early spring or later autumn, or shortened to 5 to 6 days, depending on the growth of shoots, or on purely mundane consideration like the requirements of quantity or quality of tea at a point of time. What therefore is an ideal plucking round is difficult to generalise, because as stated earlier, very many interacting phenomena are involved here. But where standard plucking is the practice, the shortest possible round could be 5 days during the monsoon , the longest being 7 days. With fine plucking the corresponding period should be 6 and 9 days respectively.

BLACK TEA PROCESSING

OVERALL STEPS

Black tea manufacturing technology involves disruption of the cellular integrity of tea shoots; in doing so, the mixing of substrates, polyphenols and the enzymes, polyphenols oxidase is facilitated. This results in the initiation of a series of biochemical and chemical reactions with the uptake of atmospheric oxygen and formation of pigmented hot water soluble polyphenolic compounds, characteristic of black tea. Black tea processing consists of the following unit operations:

- Withering
- Rolling/disruption
- Oxidation/Fermentation (biochemical reactions in the presence of oxygen)
- Drying
- Sorting (fibre removal; grading based on size) Of these, chemical changes occur primarily during withering, fermentation and drying.

Withering

Withering is carried out to prepare the green tea shoots bio-chemically and physically for subsequent manufacturing operations. Withering essentially consists of storage of green shoots for about 12 to 20 hours with the partial removal of moisture from the leaf. Withering is accompanied by certain chemical changes which may affect fermentation and thereby, the quality of the final product. Chemical wither is essentially a timetemperature dependent process. Some of these changes are dependent on moisture loss while others are independent. The major physical change during withering is the increase in cell membrane permeability and consequently, the leaf becomes flaccid. The increased permeability of the membrane has a great effect on the mixing of substrates and enzymes during fermentation. Withering is also accompanied by the activation of oxidative, hydrolytic and proteolytic enzymes, causing significant changes in the chemical composition of tea shoot, as a result the levels of soluble protein, free amino acids and simple sugars increase in the withered leaf. Amino acids increase due to the breakdown of proteins by the enzyme pepsidase. Amino acids are primarily the precursors of the aromatic compounds.

Caffeine content increases with withering. The increase in caffeine content is greater if the withering is accompanied by loss of moisture. Caffeine forms a complex with theaflavins thearubigins and "Creams" down from the liquor; this creaming down is a desirable cup character for black tea. The Chemical wither also increases the level of organic acids and improves polyphenol oxidase (PPO) activity. The increase in organic acids will have an obvious effect of reducing pH and TF formation is increased as pH tends towards 5.0. The effect of increased PPO activity is beneficial for more TF formation provided oxygen availability is not limited during catechin oxidation. Volatile compounds essentially impart aroma to the finished product. Volatile compounds present in the tea are classified into two groups : Group-I compounds, which are deleterious to the tea quality and Group-II compounds, which impart sweet flowery aroma to tea and its presence is highly desirable. The ratio of Group-II compounds to Group-I compounds is known as the flavour index and it is a semi-quantitative method of describing the aroma quality of tea. Tea with higher values of flavour index normally have better aroma quality. Withering improves flavour index and thereby, sweet aroma.

Leaf Disruption

This stage in processing is described in a number of ways, e.g. rolling, cutting, crushing, tearing, but the basic requirements are size reduction with a degree of cell disruption to allow the exposure of the new surfaces to air in the subsequent fermentation stage. The LTP machine consists of a central rotor with a number of knives and beaters, where the leaf is bruised and cut by the rotating knives and beaters. The leaf next enter into the rolling section. The rolling is done either by orthodox rollers or by a combination of orthodox rolling and the CTC machines. The surface of each roller is made up of a number of segments made of stainless steel. The withered leaf passes through a battery of three to five such CTC machines which cause the leaf to get crushed, torn and curled.

Fermentation

Tea fermentation is essentially the oxidation of simple substrates into complex characteristic substances by endogenous enzymes present in the tea leaf. This distinguishes tea manufacture from other food processes in which exogenous fermentation inducing agents are added to the raw material. Fermentation is not confined only to the period during which the leaf lies in the fermenting drums, racks or floors. It commences from the time the cells are bruised and extends until the enzymes are deactivated in the drier. As a first step during fermentation, the catechins are oxidized to highly reactive transient orthoquinones by polyphenol oxidase. The quinones in turn dimerize to produce theaflavins (TF), which are orange red substances that contribute significantly to astringency, briskness, brightness and the colour of tea beverage. Theaflavins comprise 0.3 to 2.0 percent of the dry weight of black tea. Further transformations of dicatechins and theaflavins yield compounds which are known as thearubigins, comprising about 9 to 19 percent of black tea, are red brown in colour and contribute to colour, strength and mouth feel of liquor. While theaflavin content of tea increases during fermentation and starts declining after reaching a peak thearubigin content continually increases throughout the fermentation. In actual practices, the completion of fermentation is judged by the change in colour and the pleasant aroma that develops. However, chemically optimum fermentation may be assessed by monitoring the profile of TF content and taking the time required for the production of maximum. TF Proper balance between theaflavins and thearubigins is also essential for a good cup of tea. Like any other enzymic reaction, the rate of fermentation primarily depends upon the concentration and relative proportion of individual substrates, availability of oxygen, the activity of enzyme, pH and the temperature.

Since it is not a homogeneous system, the degree of contact between the enzyme and substrate will be important. In addition, mass transfer of air to the reaction side is crucial in controlling the rate and product profile of fermentation. During fermentation, the PPO activity is known to decrease due to the formation of insoluble complexes of polyphenol oxidation products with the enzyme protein by feed back action. This precipitation process increases as fermentation continues and the temperature of the fermentation leaf increases. Both reaction rate and the ultimate concentration of products are affected by the concentration of oxygen. Theaflavin production will be inhibited if the concentration of the oxygen in the air falls below normal, which can occur in dead spots' in a forced-air fermentation system. By increasing the oxygen concentration of the air in a fermentation unit, it is possible to achieve higher theaflavin levels.

Temperature is one of the most important factors influencing the complex series of enzymic and chemical reactions during fermentation. Temperature not only affects the rate of fermentation but also the ultimate level of theaflavin in tea.

As a rule of thumb for altering the fermentation times to suit ambient temperature conditions, the rate of fermentation at any temperature will be taken as 1.7 to 1.8 times as fast as that a 10°C cooler. Though, increase in temperature accelerates the fermentation rate, maximum theaflavin occurs, when fermented at low temperature for a long time. This happens because as the temperature increases, the break-downs reactions become more rapid than the theaflavin formation reaction, and ultimately less theaflavin is produced. Another advantage of low temperature fermentation is that the peak of TF production is maintained for a longer period. Even a delay in firing results in a detectable difference in quality.

Drying

The drying of fermented tea has three major objectives; to terminate the biochemical functions by heat denaturation of the enzyme; to reduce the moisture to increase the shelf stability of black tea and finally, to enhance chemical reactions responsible for black tea character and flavour. The polyphenol oxidase enzymes which convert catechins into theaflavin are not inactivated as soon as the leaf enters the drier. In fact the fermentation process is actually accelerated and continues at a faster rate till the temperature of leaf reaches 55°C at which the enzyme is completely inactivated. Hence, any unoxidised catechin will continue to be converted into theaflavin until oxidase enzyme has been inactivated. It has been reported that 10 to 15 percent of theaflavin content in black tea is formed during the first 10 minutes of drying. A noticeable effect of drying is the change in colour of dried leaf brought about by the transformation of chlorophyll into pheophytin, which imparts the desired black colour. During drying, the reduction in stringency of fermented tea occurs due to the combination of polyphenols with tea leaf proteins at the elevated drying temperature. Drying causes an overall reduction in the quantity of volatile compounds, although certain aromatic compounds continue to be formed. The fermented Tea leaf is dried by exposing to a draft of hot air; the process takes about 30 minutes. Some of the flavour characteristic of the final product are developed as the leaf heats up during the drying process, and eventually the leaf temperature and moisture levels bring about the enzyme destruction or inactivation necessary for preservation of the product.

The conventional tea dryer is basically a rectangular chamber. The tea to be dried is spread on perforated trays. These dryers are driven by trays-carrying chains and sprockets running on angle iron runners. The movement of the trays with the leaf is counter-current to the draft of hot air. Recently the technology of fluid-bed drying at 125°C for 20 minutes is being followed in many cases. Normally the leaf enters the driyer with 60-72% moisture, depending on the cutting process used, and is dried down to 2.5-3.5% moisture.

Cooling , Sorting and Grading

The hot dried tea is cooled on the factory floor before it is sent for sorting and grading. Sorting is the operation in which tea particles of the bulk are separated into various grades of different sizes and forms confirming to trade requirements. In other words, it basically converts the bulk into finished products. The process of sorting has two objectives to impart quality. Sorting enhances the appearance and quality of liquor; at the same time it can also deteriorate the quality. The presence of fibre or flakes of coarse leaf in a primary grade causes harshness and their removal makes the liquor mellow. The cleaning of fibre also improves the black appearance of tea which is desirable. Bloom is indicative of liquor character; over sorting and over cleaning can result in loss of bloom. Usually a tea which has not been well fired, loses bloom more quickly. If tea absorbs moisture during the cleaning process, liquors can deteriorate and its keeping quality reduces. Sorting of bulk has to be done in three stages – (1) Cleaning of fibre Winnowing. The initial separation of leaf and dust can be carried out on a Middleton fitted with a sieve on No.24 mesh on the top and No.10 mesh on the bottom. Tea particles which pass through mesh No.24 are collected separately and graded further in the dust grade sorter. Similarly, particles which are retained on No. 24 mesh and pass through mesh No.10 are graded separately in the leaf grade sorter. Particles retained over 10 are generally taken for reconditioning. Both leaf and dust grades are further fractionated either in Trinic sorters or vibro sorters. After sifting, each grade is finally cleaned by winnowing during which fibres and flaky teas are removed. Generally, Super Fine Dust grades are cleaned separately in the fibrex after winnowing.

Packing

Sorted and graded black tea is normally packed in a ply wood tea chest. The chests are assembled using various components, like the plywood shooks, battens, metal fittings, iron nails and aluminium foil with tissue paper lining. Recently attention has been drawn to the use of non-conventional materials like paper, multiwall paper sacks and polythene lined jute bags.

Fuel and Power Requirements

Factories require a power source for driving machinery and a fuel source for heat generation. The nature and combination of fuel and power sources varies considerably depending on the natural resources of the tea growing area concerned, and on the economics applying to possible alternatives. The efficiency of the drying system and local meteorological conditions will also effect the fuel consumption.

Electrical Power Source

The simplest source of power is in areas where a national supply and distribution network is available. However tea estates often install their own generation facilities for running the plant and other facilities during power shutdown period. Nowadays most standby generators are driven by diesel engines, producing about 3 kWh per litre of fuel consumed. To produce one tonne of tea normally requires between 430 and 490 kWh.

PEST ATTACK AND CROP LOSSES

Tea provides a stable microclimate and a continuous supply of food for rapid build up of phytophagous species that include insect, mites and eelworms. More than 300 species of insects, mites and eelworms are active in tea areas, though not all at the same time. The simultaneous presence of different species of mites and insects, each with their characteristic mode of feeding, and diverse habitat and seasonal cycles, calls for optimal management of the pests which should be both ecologically and economically sound. A major aspect of tea pest management is the practice of simultaneous application of species specific acaricides and insecticides against pests with diverse temporal distribution with the plant itself. It is essential that these pesticides do not leave any residue in tea to avoid consumer hazards. A loss of 10 percent due to overall pest attack is a generally accepted figure though it could be 40 per cent in devastating attacks by defoliators.

PESTICIDE APPLICATIONS AND THEIR ENVIRONMENTAL ASPECTS

STEPS FOR EFFECTIVE PEST CONTROL

The tea processors should take following corrective measures during pesticides applications:

- When infested by mites occuring normally from foci, chemicals are applied and such infestation are identified in the field & properly marked.
- Chemicals of approved grade as laid down in guidelines by Tea Research Institute are only used, within the expiry date.
- Application methods normally follow the recommendation on the dosage & the recommended safety measures.
- Spray are made always after a plucking. Spraying are not overdue between the plucking intervals as more sprays often cause building-up of residue. Functional sprayers with proper functional nozzles are normally used.
- Chemicals are used on a rotational basis. Chemicals are normally not repeated more than two consecutive rounds.
- CDA applicators are normally not used.
- Correct spraying intervals are normally followed.

ENVIRONMENTAL ASPECTS

The problem of pesticide residues on made tea is receiving attention. The pesticides on tea are generally used after plucking. If the sprays are done between pluckings the residues can be high. As spraying is done after plucking the deposits are left on the mature leaves and on shoots and leaves which are in different stages of growth.

The weight of these immature shoots increases over a period of 7-14 days depending on the plucking intervals. Therefore, the pesticides undergo what is called a growth dilution, which should not be mistaken for the pesticide degradation. By the time immature buds attain the size of pluckable shoots the residues of the pesticide on them will be very low. It is estimated that the weight of all shoots increases approximately 13 times in a week and 80 times in two weeks. Pesticides applied on tea are degraded and also further diluted by rain and dew, photolysis through sunlight and biodegradation. Compounds like dimethoate, fenitrothion, and quinalphos have all half-life below two days, though for ethion the value is 3.4 days. Dicofol has a half-life of nearly four days. Sensitivity of pesticides to light is an important factor which determines their residues on tea. The mechanism for most photochemical reactions involves the photolytic cleavage of covalent bonds. Since these reactions are based on functional groups, it makes no difference whether a pesticide is an insecticide, fungicide, or herbicide. Under experimental conditions the photodegradation of some of the commonly used pesticides is as follows:

Quinalphos > Malathion > Chlorpyriphos > Fenthion > Fenitrothion >Dimethoate > Phosalone > Ethion > Aldrin > Dicofol > Deltamethrin >Fenvalerate > Cis-permethrin > Cis-cypermethrin.

Considerable amount of pesticide on the leaves is lost during the process of manufacture. The loss may be 30-60 per cent. Degradation in this stage may be due to evaporation and thermal decomposition. Compounds with higher vapour pressure are likely to leave less residues than those with lower vapour pressure.

Processed tea when infused with boiling water will extract small quantities of residues in tea liquor. Most of the pesticides used in tea have very negligible solubility in water. In fact, it is the liquor which is consumed and not the tea as such. Based on the degradation rates of the chemicals that are used in tea, the safety intervals can be classified as follows :

- Pesticides with rapid degradation rate. In these cases the safety interval is set at five days. Dichlorvos (DDVP) and the pyrethroids.
- With moderate degradation rates, the safety interval is set at 6-10 days; Malathion, Fenitrothion, Quinalphos, Dimethoate, Phosphamidon, Phosalone.
- Pesticides with slow degradation rate where the safety interval should be more than 10 days: Ethio, parathion, Methyl parathion, Demeton, Tetradifon, Dicofol, Chlordimefom.

ENVIRONMENTAL MANAGEMENT PLAN

As stated earlier the tea industry does not pose any threat to the environment. The industry does not generate much waste. It has been seen that a significant portion of the waste are being re-used in some way or the other. However, the industry should take up certain measures/actions, which may be useful in improving & maintain the environmental performance of the industry.

AIR POLLUTION CONTROL

Source of air emission in the tea industry is basically the D.G. set, which is operated only during the power shutdown period. The following efforts are needed from the management towards controlling the air emission.

- Use of low sulphur fuel for combustion purpose in D.G. sets.
- Use of tall stack.
- Avoiding frequent start-up and shutdown of D.G. sets.

WASTEWATER MANAGEMENT

Source of wastewater in the tea industry is the domestic sector since the industrial process is basically dry in nature. Hence it is recommended that the domestic waste thus generated be treated in simple treatment system prior to its final disposal. Infact, the wastewater so generated may be treated in Septic Tank. Partially treated effluent from the septic tank should either be disposed off in soak pits or to be further treated in up-flow anaerobic filter (UFAF). Soak pits are cheap to construct and easy to operate. These pits need no media when lined or filled with rubble or brickbats. When water table is sufficiently below ground level, soak pits should be preferred only when land is limited or when a porous layer underlies an impervious layer at the top, which permits easier vertical downward flow than horizontal spread out. Minimum horizontal dimension of soak pit should be 1 m.

The pits should be covered and the top raised above the adjacent ground to prevent damage by flooding. In such case where suitable conditions do not exist for adoption of soil absorption system, the partially treated effluent from the septic tank should be treated in up-flow anaerobic filter (UFAF). The up-flow filter can be successfully used for secondary treatment of septic tank effluent in areas where dense soil conditions, high water table and limited availability of land preclude soil absorption or the leaching system for effluent disposal. It is a submerged filter with stone media. The septic tank effluent is introduced from the bottom. The microbial growth is retained on the stone media making possible higher loading rates and efficient digestion. BOD removals of 70% can be expected. The effluent is clear and free from odour. This unit has several advantages viz. (a) a high degree of stabilization; (b) little sludge production; (c) low capital and operating cost; and (d) low loss of head in the filter in normal operation. The up-flow anaerobic filter can either be a separate unit or constructed as an extended part of septic tanks. The treated effluent from the UFAF could be discharged into any inland surface water or can be used for gardening purpose. Apart from such treatment the management should also practice the following :

- Reduce wastage of water.
- Reduce volume of water use.
- Use of treated water for gardening purpose.

SOLID WASTE MANAGEMENT

The tea plants provide some green manure from leaf-fall; where shade trees are planted their leaves also form green manure. The waste from factories amounts to about 2% of total production. Currently it is either returned to the field as fertilizer or sold at a very low price. Products which can be produced from tea waste are shown below :

- Caffeine
- Polyphenols
- Pigments
- Polymers (tea polyphenols can be used to replace some of the phenol in phenol formaldehyde resins)
- Animal feeds
- Foaming agent
- Vinegar
- Tea seed oil (alternative to groundnut and olive oils for cooking, cosmetics and pharmaceuticals)
- Furfural.

OVERALL MANAGEMENT

The overall measures/actions that the industry may take up with an aim to improve the environmental performance of the industry are as follows:

- Practice Good House Keeping.
- Maximum re-circulation / re-use of the wastes.
- Conservation of domestic water.
- Quality control laboratory should monitor the residence of pesticides in finished tea.
- Generally finished tea is packed in poly-bags and these poly-bags are placed in jute bags for final delivery. It is suggested that the plastics used for such packaging should be of food grade in nature.
- Tea industries may try to use food grade jute bags instead of odourless jute bags as are currently being used.
- Attempts should be made to use organic fertilizers and pesticides in place of chemical fertilizers and pesticides.
- Attempt should also be made to optimize the use fertilizers and pesticides during tea production.
- Periodic monitoring of a few selected water sources around the tea garden should be made.
- Periodic monitoring of health condition of workers both in garden and factory should be made.

MINIMAL NATIONAL STANDARDS

The prime concern today is to prevent pollution of the physical environment of air, water and land such that the entire living system including the plant is not adversely affected. Indiscriminate pollution of the environment in the past has seriously affected the plant kingdom, animals and birds. Human beings are the worst victim, who are directly affected by the pollutants as well as by the loss of animals, birds and plants. There is an urgency for the abatement of pollution and it is essential that no further damage is done to the environment. Industries are one of the major contributors to the pollution.

It is therefore, necessary that Rules & Regulations are laid down to avoid further deterioration of the environment. The Central Pollution Control Board (CPCB) has been entrusted to frame and implement Rules & Regulations in the country to prevent any further pollution as also control such pollution which are already being contributed by the industries. The Central Board operates through similar Boards at the State level who have the statutory powers to lay down the Rules & Regulations to control the indiscriminate pollution of the environment in their respective states.

Pollutants which are being discharged by the industries differ in their characteristics. It is therefore, necessary that standards be laid down industry-wise for the permissible levels of pollutants that can be let out by them to the environment. Such standards are available now for most of the major industries of the country. The Tea Processing Industry however is one of the exceptions in this respect. It is in this context that the CPCB desires to formulate 'Minimal National Standards' in respect of the liquid wastes and emissions for the Tea Processing Industry.

It has been observed that the tea industry is a dry industry. However, liquid waste of domestic nature is generated in the township and labour lines of the gardens. A small amount of liquid waste is also generated from the domestic sector in the industry. There is no source of air pollution in the tea processing. However D.G. sets are installed in a few gardens which are operated only during the power shut-down period. The solid waste from a tea industry is either used as fertilizer or used for making other by products. Considering the above the following standards are recommended for the consideration of CPCB.

ELEMENTS VITAL TO 'QUALITY' MANUFACTURE

The work that is involved on an estate to ensure that healthy leaves continue to flourish and are skillfully prepared for your comforting cuppa!

THE BUSH

In its normal state of growth tea can grow into a tree of some 35 feet in height. For easy control and constant harvesting, planters have developed a pruning and plucking system which controls the bush at about two and a half feet in height, encouraging it to spread and fill across to the surrounding bushes. We call it the 'plucking table'. Below are listed some of the main works to invigorate and nurture this amazing plant:

Pruning

Each field is maintained on a cycle dependent on elevation.

- Low elevation - pruned every 2.5 to 3 years.
- Mid elevation 3 years.
- High elevation 4-5 years.

Each year about 2 fields in each division will need pruning and brought back bursting with energy, to go back into plucking. A division is usually an area of some 200 to 300 acres. The area is divided up into fields, often with no obvious boundary. These fields vary between 20 to 50 acres each and about 7 - 10 fields per division. When the first tracts of land were purchased from the crown for clearing, purchases were often in multiples of 50 or 100 acres. Pruning these days is fairly light in the main, it is largely a slash pruning, years ago we would have religiously cut out all dead wood back to live wood and shortened all new shoots to 4 inches. Recovery of the shoots back to tipping is about 90 to 105 days for new shoots to reach the level of the plucking table. Once the long green shoots are up tipping takes place. This is the first plucking in a new field from pruning and it breaks the shoot back to where the plucking level is required and this will be the plucking table until the bush is pruned again. The broken and plucked shoots are thrown on the ground as they are useless for quality tea and rolling. From then on the plucking cycle begins again every 7 days.

Each division has a band of trained skilled pruners and trains up two or three new pruners each year. Each pruner is given a set number of bushes 140 - 150 for a days work.

Drains

At the time of pruning the field is spring cleaned. The hillsides are criss crossed with boxed drains on the contour, some 2 feet deep, with a small dam every three or four feet. This retains the heavy monsoon rain with any soil that is eroded and allows the soil to settle, feeding the water on over the dam to the next box and so on. The drains feed into the deep natural ravines. At pruning time the drain boxes are re-cut and the soil returned and spread amongst the tea bushes above the drain again.

Weeding

This is a monthly exercise which is carried out fairly gently these days without scraping the soil to any extent. Grasses and certain binding weeds are encouraged to combat soil erosion these days whereas in earlier years weeds were eliminated. Traditionally every family received an area to work in their spare time for increased income. Usually the family would then leave grandmothers and grandfathers to do this, together with younger children who enjoy being in the fields. Any members of families who may have some genetic or physical problem that does not allow them to work in the normal sense could also be given weeding work which is light and provided an income to the family unit.

Manuring

In my grandfather's days he often used horse and cattle manure, with compost and Poonac (the by products from the coconut mills) for manuring. Later the Colombo Commercial Co and Baurs introduced the modern scientifically formulated artificial manures and a formula was born for calculating the requirements of the bush. This is based on the application of 10 lbs of Nitrogen for every 100 lbs of made tea yielded. This and increasing areas of the new high yielding Vegetative Propagated (VP) bushes, resulted in rising yields and capacity problems in withering, which led to the trough withering system. Certain problems have arisen. For instance the 2025 clone combined with heavy manuring produces a very fleshy, watery leaf and this has created some problems when trying to maximize quality. The trough withering system is now the standard system and is one of the few new mechanical advances that have truly been as good as or bettered the old system.

Disease

In the 1960's spraying of chemicals was the standard response to disease. D.D. for root disease fumigation and Dieldrin for shot-hole borer beetle, copper spraying for Blister Blight fungal spore attack on the leaf. It was realized that Dieldrin created as many problems as it solved, by killing the Braconid Wasp which parasitized the Tortrix moth and then led to Tortrix attack. The TRI concentrated on biological control, encouraging the right species of insect for nature to do the controlling. This has been a huge success story in the island and the International ISO organization recently congratulated the industry on its performance and the production of the cleanest teas in the world. Spraying residue levels were not recordable.

Nurseries

These are vital to sustain large supplies of stock for speedy planting in the short planting season, as the monsoons break.

Tea Nursery

A tea nursery, consists of a plot of mother bushes, of the clones that the estate has selected for their purposes. These mother bushes are allowed to grow their new shoots naturally until they reach about 18 inches in height when they are carefully cut with a very sharp knife by the nurseryman and he then proceeds to cut above each leaf node to provide a leaf with a short stalk beneath. These nodal cuttings are carefully retained in a bowl of water and then planted into ready made large beds of 1,000 poythene tubes filled with a prepared soil mixture. Originally the beds were stocked with bamboo baskets but to-day polythene tube is cut for the plant bags. The beds are shaded from the sun, watered and fed regularly by the nursery staff. Planted out at a year old and then cut across twice a year for a year to 18 months to encourage width before being brought into yield at 2 to 2.5 years of age.

Each mother bush can supply some 1,000 cuttings per year and are pruned at 15 to 18 months to encourage further supplies. The original areas planted mainly in seed, are now 125- 130 years old and still in production but failing. This has meant a large replanting programme from the 1950's accelerating in the 1960's and still underway. The costs of this are very large indeed and a huge drain on estate finances, even with government subsidies. The basic premise is every estate should re-plant 2 % per annum of the total estate acerage in order to keep to an on going cycle of renewal that will maintain the ideal yield and therefore keep overhead costs down to a sensible level. As soon as yield falls, the cost of production of 1 kilo of tea rises.

Present day thoughts on the life of the bush have been revised in line with the greater yields and stress on bushes compared to the 1900's:

- Seed sown tea = 75 - 80 years life
- V.P. planted tea = Ideally Low country 35 years. High grown 50 - 60 years.

When new planting is undertaken. the whole area is cleared and winched out. The soil is then deep forked down to 18" by hand, removing all roots, leveling, terracing and planting in Guatemala grass to renovate and improve the soil. This grass is lopped twice a year and laid on the soil to mulch for a period of two years. The area is then holed and planted with the one year old plants at a rate of 13,500 per hectare. Planting distances have varied over the years with 4 feet × 4 feet, 4 × 3 feet and then 4 × 2 feet.

Shade Tree and Firewood Nursery

This is a separate nursery maintained to bring on species of trees for planting within the tea as shade trees , for wind belts and firewood plantations (gum). Some estates also plant up copses of pine and jungle trees for timber production. You will see that we use an estate tree (sustainable under this programme), for our wooden chestlets or boxes. At Le Vallon there was a green manure nursery in Colgrain division above Colgrain bungalow and I then opened a specialist gum nursery in No.2 field alongside and above the main cart road. At that time there was a large expanse of patna grassland above the main Government road at the end of Colgrain division which had once been a coffee estate with a tumbled down stone bungalow covered in bushes and small trees. The area was well over 100 acres in extent and a huge flow of blue gum plants were required to plant this up with fire belts cut across the hillside. This was to become a valuable source of renewable firewood for the two tea factories on Le Vallon as blue gum once felled will re-generate itself like coppicing. Thus the estates in those days were always thinking of ways that would allow the estate to be as self sufficient as possible, cutting down on expensive imports of oil to fire the furnaces.

Many factories were designed to take alternative fuels. Le Vallon factory was able to take firewood for its burners to fire the leaf, plus fuel oil and pelton power for driving the belts to drive machines and produce our own electricity. Today large internationally financed hydro schemes are in place. In the 1940's our own family estate was totally dependent for electricity produced on the estate. This meant the charging of large banks of batteries in the bungalow during the day, ready for use at night. Each night the factory requested permission to stop the machines and the bungalow batteries took over. If the bungalow was hosting a party then the factory would be requested to keep the engines running and the night watchman would receive a telephone call from the bungalow when the engines could stop.

Plucking

One of the most important operations to the tea-maker. I have discussed pruning as the starting point and the fact that from pruning the bush takes about 100 days to recover growth to the plucking table.

Each division has a new field plucking force, of usually 60 highly skilled pluckers . They move in on the designated day with long level sticks and break back the new flush to the level, following every change in the contour of the ground, this gives an amazing smooth rolling carpet effect across every hillside and the new field will be a light fresh green colour. This first break is thrown to the floor in quality production and then the field comes into the divisional plucking cycle with every bush being plucked every 7 - 10 days depending on the weather.

7 days in wet weather, 10 days in arid dry periods, every day of the year. The pluckers fingers flick across the bush selecting the fresh new shoots with a bud and two leaves, some estates will take four leaves. Traditional rolled leaf requires greater disciplined plucking. CTC can deal with much coarser leaf. The pluckers also break back any coarse leaf or dormant shoots and throw them down on the ground, moving to the next bush and so on. Today it appears that the plucking round is set in stone at 7 days in rain or drought. The tight round apparently increases yield even if there is less leaf to take.

A new field is designated as new for up to 2 years and then handed over to the old field pluckers who are not as skilled. The new field pluckers are numbered 1 - 60 and allocated rows throughout every field and they know those rows belong to them. This is used in determining and supervising responsibility for poor plucking. Once a field becomes an old field on its way to the next pruning, it produces a more mature leaf and the tea-maker will be aware of which fields are coming in each day.

Their age and aspect on the estate and he will respond in manufacture accordingly. There are many disciplines to be followed in plucking for the best production. It is vital to ensure that the leaf is weighed often enough, especially in heavy flushing periods after rain when yield increases, to ensure that the leaf is not crushed into the baskets or waiting around on the roadside for long periods. As soon as the leaf is plucked, withering starts and fermentation of any crushed leaf also starts. It is vital that estates invest in open weave baskets allowing air movement and bamboo supports to allow baskets to be stacked without crushing.

At weighing it is important that leaf is not left lying on the floor and walked on but cleared constantly. With many estates having large areas of re-planted tea in the new clones the bushes have filled out to join each other and walking between the rows is harder, this has brought in plucking into sacks and then the leaf from the sack is transferred to the cane baskets at the field edge. I understand the idea but I am not convinced that this is best for the leaf as crushing from extra handling and shaking must take place.

I also see too many instances of plucking that is too hard and includes leaf that is not ideal for rolling. I understand that these days the pluckers will down tools if they are disciplined too much because they lose weight and bonuses but tighter control would lead to better quality leaf. Green tea estates also practise this and perhaps they can get away with it where steaming is used. Certainly the modern C.T.C. system of 'Cut, Tear & Cut' machines will make a product from anything if quality is not a requirement.

Classification of Teas by Seasonality

A major concern here in the U.K. has to be the level of knowledge that consumers or indeed the general trade has about tea and how it gets to them. Once upon a time tea was part of what people learnt at school in Geography along with coffee and other products that the U.K. had been involved in during the Colonial era. One of the basic facts that gets lost in marketing spiel by large companies is the classification of a tea by when it was produced during the year. Commercial companies convey as little as possible about their teas when they market them. It is as if they wish to divorce the production process from the marketing process.

The very finest Ceylon teas will be produced at a very specific window of weather such as the Western quality teas in February to the end of March and within that window will be an even smaller window of about one to two weeks when the season is declared a season and the flavours are really concentrating within the leaves. You can have single estate teas - They have to come from a single estate but when??? A single estate can be from that core S/W or Uva season period - or just in the window either side which would be the next quality down as long as there is little rain on the estate. A single estate standard - These are still single estate but they are teas produced from the daily programme come rain or sunshine, blended up to a tea board standard laid down for each district or alternatively to a client requested standard. To bring such teas up to these marketable standards will require 8 % to perhaps 20 % of the classic seasonal quality leaf to be added to teas plucked in the monsoonal wet seasons.

Beware of seeing packets marketed with an estate name on without a reference to it being of seasonal quality production or authenticated by the area tea board. If you are on a website, they may borrow ideas from other areas such as tea board sites or industry information sites and suggest that their teas are of the highest quality but unless you can detect quality you will be unable to classify the tea. Obviously price will in the first instance generally define a standard from a seasonal tea. We buy our teas direct, we pack and ship those products direct to the U.K. & our foreign clients so even though their prices may seem reasonably high they are as low as we can possibly market them.

Yields

Early estates including our own in the 1900's were yielding 400 lbs of made tea per acre. (Remember from 100 lbs of green leaf in the field, we get approximately 20 to 22 lbs of made tea). By 1953 our estate was yielding 915 lbs per acre. By 1962 it reached 939 lbs per acre. We sold the estate in 1954 to the Thondaman family (started the first major union – CWC). The national average yield in 2005 was 1,500 lbs per acre .

Manufacture

The most important phase of all in quality production, once the disciplines have been followed in the field. I have discussed pruning and the age from pruning. The night before the receipt of the following days leaf the tea-maker will assess which fields the divisions are sending in and what yields that they anticipate. In a six division estate, he will be receiving at least 6 fields of new leaf and 6 of old field leaf.

The new fields will need a different approach to the old. He will consider whether they are new VP planted fields or old seed stock. The field work history of each of the fields will be known to the tea-maker. Whether they are Assam or China jat leaf, whether it is rainy season leaf or dry season leaf and finally whether the field is a newly planted field or perhaps just recovering from pruning. All this will determine his rolling programme and the pressures applied, plus the length of fermentation.

In the dry periods fermentation can take 35 minutes to 3 hours and in wet periods heat will be applied which will affect leaf quality. Present day policy in the factories tends to a four roll programme especially for the larger grades or a two roll programme, followed by two rotorvane passes to produce maximum levels of BOPF and BOP grades.

The roller is often a 47" roller charged with 250 Kilos of withered leaf every 55 minutes. Final capacity is determined by the drier output. The drier operates at 180 to 200 o F and arrests the fermentation process, as well as lowering moisture level ideally down to about 2 or 3 % which is ideal for keeping the tea and should never be exceeded. Over 5 % and the leaf loses its sheen and becomes dull and grey looking. Rolling produces friction which creates heat, an enemy of quality tea production. In the flavoury districts like Uva manufacture can take place after midnight to try and use lower temperature in the rolling rooms, helping the management of the volatile oils expressed in rolling. Below is a fictitious example of how temperatures rise and how the tea-maker tries to control them. Rotorvane & C.T.C. systems are more liable to create excessive frictional heat & shattering of the leaf with dusty particles.

Roll breaking is when the dhool is discharged from the roller and passed over the roll breaker which has a vibrating sieve.

Room Temperature:	**After one roll:**	**After Roll Breaking:**
60 °F	83 ° F	70 ° F

The leaf that passes through the sieve is accepted as the first dhool & sent to the fermentation process. The leaf that passes over the sieve is sent to the next roller for further conditioning.

Manufacture of traditional teas is a highly skilled operation. Tea-makers train for decades as juniors, before being promoted. Manufacture is a huge subject with varying opinions and it is only possible to give you a flavour of the subject here. Obtaining top quality teas and maximum flavour in high season is an enormous operation with close liaison between estate and ourselves. Estates normally run for maximum levels of the smaller BOPF and BOP grades in the high dry seasons to emphasize flavour. Asking for large grades in high season means special runs upsetting their normal programmes. Emphasis on FBOP grades at this time will degrade the smaller grades and prices will reflect this. Certain estates have been very co-operative even allowing special runs at night to obtain cooler conditions. Our whole ethos in working with the estates in this way is to promote their individual identity to the public, something that has been lost over the years. We believe that this connection between consumer and producer is vital for the consumer to be able to appreciate and keep in contact with the process that delivers their cup of tea, day in and day out.

Hopefully then they will begin to appreciate this truly amazing bush and the work that goes on to make 'tea'.

Manufacture has changed from the methods used up to the 1970's, largely because of the introduction of the tea bag. Prior to this, Ceylon teas satisfied the connoisseur who wanted maximum development of flavour and leaf that was pleasing to the eye

. Large leaf teas with a good black look and as much tip as possible. 1960's B.O.P. grades were marketed at 3 times the size of present day B.O.P's which are cut & cut to suit the commercial market. Below I have tried to summarise the present trend which tends to follow certain elevation criteria.

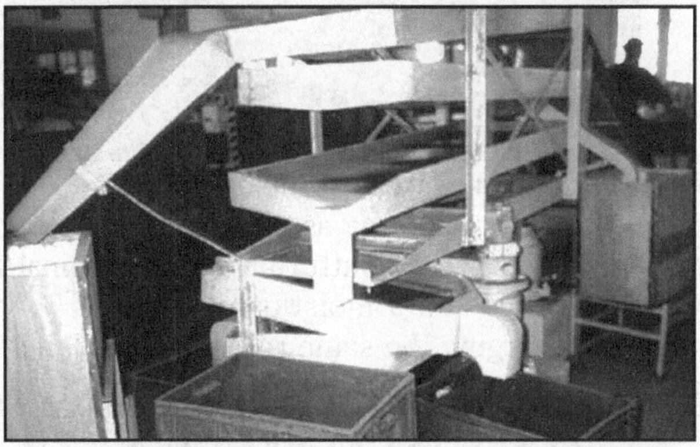

With the traditional twist roller system we were only able to turnout 1,730 lbs of made tea per day. This was achieved with a 3 roll system and the rollers could be of various types, we were using Jackson 36 inch & 28 inch rollers. On the 3 roll system we would charge 2 rollers for the first charge of 500 lbs of withered leaf into the two rollers , 2 rollers for the second rolling charge, reducing to 1 roller for each of the last rolls. Between each rolling we extract what we call a dhool which is the leaf that is dropped through the bottom of the roller table into a trolley and then passed over a roll breaker. The roll breaker is a long vibrating machine with wire mesh base that is made to different sizes, dependent on the size of leaf that one wishes to extract also allowing the leaf to cool as it is vibrated over the metal mesh sifter, the leaf that passes over the sifter is returned to the next two rollers to be rolled again, when another dhool is extracted. These dhools then become graded by leaf size. A rolling took 35 minutes to complete and after 20 minutes, the 1st. roller was cleaned & ready for action again with the start of the 2nd. programme whilst the first was progressing through the other rollers situated on the rolling room floor and this process can continue for 7 programmes in a normal day. This entails 6 programmes charging two rollers at 500 lbs between them and a final programme where 660 lbs was charged into the two rollers. These 7 programmes totalled took a total of 3,660 lbs of withered leaf turning out 1,732 lbs of made black tea.

The tea-maker would decide on a) the amount of pressure to apply to the screw and often this could be half pressure for the first roll and then full pressure for the last two rolls, however he would only apply the cap down for approx 4 minutes and then release the cap for 2 minutes to allow the leaf to churn and also most importantly cool. The churning allowed the leaf at the top to come into contact with the roller table at the bottom .

7,481 lbs of green leaf plucked from the field became 3,660 lb after withering and this became 1,732 lbs of made tea as shown above after firing the days total manufacture which gave a 23 % outturn from the original 100 % of green leaf plucked by the pluckers and a moisture loss of 77 % over the whole programme for the day, charging the same rollers 7 times in a day.

PROCESSING TECHNOLOGY AND HEALTH BENEFITS OF GREEN TEA

The tea plant belongs to the genus Camellia , a member of the family Theaceae . Primarily two basic varieties of tea has been recognized as Chinese variety , and Assamese variety . From ancient time tea was used as a beverage in China and approached towards the Europe in 16th century. In 19th century , cultivation of tea plant has been started first time in India . Tropical and subtropical climate is suitable for tea cultivation with varying range of other variables like temperature range , altitude , and soil characteristic (containing higher amount of iron and manganese, PH range between 3.3- 6.0) . Green tea is one of the most popular drinks and millions of cups were consumed every day in entire world .

PROCESSING TECHNOLOGY OF GREEN TEA

Green tea is processed in different way during manufacturing. Freshly harvested young and unfermented leaves are immediately steamed to prevent fermentation, yielding a dry, stable product for producing green tea, could allow a slight amount of indoor withering for removing the moisture . Processing technology preserve natural poly phenols with respect to the health-promoting properties and poly phenol compounds in green tea are dimerized to form a variety of theaflavins, such that these teas may have different biological activities . Generally soon after harvest and inactivated enzymes cannot break down the tea tissue chlorophyll, which contributes to the green colour. For understanding the above technology, stepwise processing has been discussed bellow:

PLUCKING

For obtaining good quality of tea the buds including the two leaves or shoots with three leaves has to be plucked and processed. Tender and uniform tea leaves commonly desired for best quality of green tea (like Longjing, Biluochun, Gyokura etc). Immature shoots and coarse leaves are unacceptable because standard flush depended upon leaves standard . Coarse leaves plucking reduced the quality of prepared tea. After plucking the shoots generally spread out in bamboo trays or on the ground for 1–3 hrs to emit some grass-like odour and reduce the moisture up to optimum level to enhance the quality of green tea.

Table. Percentage of Standard Flush for Standard leaf Plucking.

Percentage of standard flush	Leaf standard
More than 75%	Fine plucking
60-75%	Medium plucking
Less than 60%	Coarse plucking

Plucking of tea leaves are mainly performed by experienced person who can pluck around 40 kg of tea leaves per day. Manually plucking required more labour involvement, hence more costs, and also more time consumption. Therefore for an efficient, cheap and quick tea leaves harvesting, the mechanization of tea leaves harvesting is need of hour. In India tea leaf mechanization is in its infant stage . Design and development of conceptual harvesting machines has been facilitating the harvesting process of tea leaves.

WITHERING

The withering reduces approximately 30% of the humidity content of the leaf within a time of 4- 12 hours. During withering, physical (leading to a loss in moisture and changes in cell membrane permeability) and chemical (influencing the formation of the tea aroma, which is important for flavour teas) both changes was occurred . Withering can be practised by many methods but normally traditional method of withering is preferred where desired moisture is reduced through air blown method .

PHYSICAL CHANGES OF GREEN LEAF DURING THE WITHERING

Fresh leaf starts to lose moisture immediately after plucking. In withering process the removal of water occurred with faster rate through the lower surface when stomata of the lower leaf surface begin to close compared to the upper surface which has no stomata . As the degree of wither progresses, the permeability of the cell membranes in tea shoots increases.

BIOCHEMICAL CHANGES IN THE GREEN LEAF DURING WITHERING

Several chemical changes occurred during withering operation has been observed . Reported biochemical changes in withering are the conversion of proteins into amino acids, reduction of lipids and fatty acids, carotenoids and chlorophylls, increment in caffeine content and changes in sugars, organic acids, PPO activity and aromavolatile components .

FIXING

After plucking the next step is fixing of green leaves in processing technology of green tea. Fixing operation is performed for reducing or stopping the enzymatic activities in green leaf for preventing the fermentation and retains the green colour of the leaves. Temperature Applied for pan-fixing is more than 180°C but in the case of steam-fixing, temperature is usually 100°C. Leaves are moved through a rotating drum and hot steam is added up to 2 minutes then leaves are extracted again. The amount of steam is the deciding factor in this step because too much amount can spoils the leaves and too little initiates the onset of the fermentation. Steam process is beneficial in retaining the antioxidant components, appearance and texture of the cooked product, decreasing the process time .

Steaming process caused the inactivation of enzymatic activity in freshly picked tea leaves and decreased the oxidation . Water extracts from steaming green tea contained 22-33% phenols and the flavanols along with catechins accounted 49-87% of total phenols. The phenolic compounds which responsible for effective antioxidant properties usually are the most abundant water-soluble components in the tea .

FIRST DRYING

Wooden or metal drum can be used for drying in which leaves are swivelled for nearly 30 minutes at 55°C warm air. During this process, the leaves lose about 50% of their rest moisture content. Physical and chemical changes occurred during drying process affect the final tea quality and that is why control on proper drying process is necessary to preserve quality and minimize energy inputs. Loss of quality and burnt taste is controlled by regulating the drying temperature . Drying can be performed with several methods like Hot-air drying with a blast oven, vacuum drying with blast oven and vacuum pump and Microwave drying and microwave vacuum drying with microwave furnace with/ without vacuum. When microwave energy has been applied to tea, water and other polar molecules of tea are induced for simultaneous high-speed rotation due to microwave irradiation, leading to the surface and interior heating at the same time, and resulting in a large number of water molecules escaped from the tea . It can be seen that rehydration ratios varied from different drying methods microwave vacuum drying>microwave drying>vacuum drying>hot-air drying, shows the maximum and minimum rehydration capacity of green tea for microwave vacuum drying and hot air drying respectively. Combination of short drying time and low temperature has least damage on the cell structure of green tea in microwave vacuum drying. Maximum nutrients (like amino acid, protein and caffeine) were retained in green tea by microwave vacuum drying and least by hot air drying. .

ROLLING

In this processing step, leaves are rolled in a rolling machine for about 10 minutes with differing pressure. Leaves passed through the cleaner to separate the impurities before going to the rolling machine after that crush particles are passed through Rotor vane machine for further crushing which again passed through Curl-turn-cut machine to make the crushed particles finer and finally materials are passed through a roll breaker for breaking the twisted balls caused slow fermentation process. Optimum rolling is necessary because of less rolling results uneven crush particles and more rolling causes loss of chemicals results, improper mixing of chemical with enzyme .

Loosen the compressed or compact leaves with the help of roll breaking followed by rolling process for drying with faster rate. For betterment of final product, removal of moisture and enhancement in flavour, drying is usually repeated several times. Steamed green tea was processed with fixation of green leaves by steaming the leaves kept in perforated drums supplied with a steam blast for a few minutes. Steamed green teas are generally needle-shaped, with dark green dry leaf; green infused leaf and typical aroma also contained vitamin C more than black and Oolong teas due to the oxidation during the fermentation process .

FINAL DRYING, ROLLING AND POLISHING

Further drying operation follows in which leaves are kept in contact with hot air up to 30 minutes for further drying after that final rolling step has been taken into account by keeping tea leaves between two rotating metal plates of rolling machine up to 15 minutes then finally leaves are polished with help of polisher by pressing the leaves against a hot plate. This makes the leaves very flat and glowing. For aesthetic point of view this step is important.

FACTORS AFFECTING GREEN TEA QUALITY

For partially fermented tea, quality is more important than yield. Environments: Soils and climate are two major factors affecting the quality. High elevations and cultural practices (including tillage, weeding, fertility management, irrigation, plant protection and harvesting) management may affect the quality of tea. Several mentioned factors affect the quality of green tea like processing technology (withering, shaking, panning, rolling, and drying), tea plant cultivars, tea shapes, fertilization and plucking methods etc. . The height of tea plant, number of leaves and duration of harvest vary with weather conditions for a particular type of green tea. Heavier fixing required to the tender leaves but mature leaves need lighter fixing and fixing temperature should be higher in beginning then reduced further are a general perceptions of china. Tender leaves contained higher enzyme activities than the matured leaves so that for deactivating the enzymes, high temperature and long fixing time is required. Protein hydrolysis rate enhanced through heavier fixing of tender leaves (like amino acid contents increased with increased fixing time) . Mature leaves require lower temperature in comparison of tender leaves for fixing because mature leaves have lower water content and enzyme activities than tender leaves. Proper shading and adequate sun light is necessary for producing high quality green tea . Nitrogen should be applied for improving the quality of green tea by enhancing the amino acid content. Based on fermentation, tea has been categorised as green , oolong (obtained after partial fermentation) and black tea. Commonly, It has been noticed that green tea is better than other types of tea which contain (–)- epigallocatechin gallate .

HEALTH BENEFITS

Regular consumption of green tea imposed several benefits on human health and prevents the human body from different disease (like Cancer, leukaemia, liver disease, diabetes) . Disease related to teeth caused by several different reasons of nutrition and bacterial infections also decreased by the green tea consumption . Green tea contains anti viral properties due to the antioxidant characteristics of poly phenols and enhances the enzymatic activity which can reduce the damage of cellular membranes and prevents the damage of cells and damages the viruses which cause the oral cavity . Drinking green tea and/ or ingesting green tea bioactive compounds may mitigate bone loss in elderly women and men, thereby decreasing their risk of osteoporotic fractures . The tea leaves also contain caffeine which has a pleasant, stimulating action and contained some cancer-preventing elements. People with difficulty sleeping may use a decaffeinated tea. Further more research has explored that the green tea consumers have lower risk of heart related disease and cancer . Intensive and broad studies have been conducted to observe the effects of tea consumption on severe disease and found that green tea consumption reduced the cancer .

Consumption of green tea reduced the cardiovascular risk and also helps in maintaining the blood pressure . Green tea contents active constituent, (2)-epigallocatechin-3-gallate (EGCG), which caused the prevention and treatment of several diseases such as cancer . Consumption of green tea was also enhancing the antioxidant activity in the blood and exerts the good effects on the brain with the help of antioxidant activity. Green tea leaves contained L-theanine caused the "umami" and posses many beneficial health effects (like improved immune function, lower blood pressure and reduced liver injury) .

5

Tea Grading

In the tea industry, tea leaf grading is the process of evaluating products based on the quality and condition of the tea leaves themselves. The highest grades are referred to as "orange pekoe", and the lowest as "fannings" or "dust". Pekoe tea grades are classified into various qualities, each determined by how many of the adjacent young leaves (two, one, or none) were picked along with the leaf buds. Top-quality pekoe grades consist of only the leaf buds, which are picked using the balls of the fingertips. Fingernails and mechanical tools are not used to avoid bruising. When crushed to make bagged teas, the tea is referred to as "broken", as in "broken orange pekoe" (BOP). These lower grades include fannings and dust, which are tiny remnants created in the sorting and crushing processes. Orange pekoe is referred to as "OP". The grading scheme also contains categories higher than OP, which are determined primarily by leaf wholeness and size. Broken, fannings and dust orthodox teas have slightly different grades. CTC teas, which consist of leaves mechanically rendered to uniform fannings, have yet another grading system.

MANUFACTURE AND GRADES

ORANGE PEKOE

Orange pekoe also spelled pecco, or OP is a term used in the Western tea trade to describe a particular genre of black teas (orange pekoe grading). Despite a purported Chinese origin, these grading terms are typically used for teas from Sri Lanka, India and countries other than China; they are not generally known within Chinese-speaking countries. The grading system is based upon the size of processed and dried black tea leaves.

The tea industry uses the term orange pekoe to describe a basic, medium-grade black tea consisting of many whole tea leaves of a specific size; however, it is popular in some regions (such as North America) to use the term as a description of any generic black tea (though it is often described to the consumer as a specific variety of black tea). Within this system, the teas that receive the highest grades are obtained from new flushes (pickings). This includes the terminal leaf bud along with a few of the youngest leaves. Grading is based on the 'size' of the individual leaves and flushes, which is determined by their ability to fall through the screens of special meshes ranging from 8–30 mesh. This also determines the 'wholeness', or level of breakage, of each leaf, which is also part of the grading system. Although these are not the only factors used to determine quality, the size and wholeness of the leaves will have the greatest influence on the taste, clarity, and brewing time of the tea.

When used outside the context of black-tea grading, the term "pekoe" (or, occasionally, orange pekoe) describes the unopened terminal leaf bud (tips) in tea flushes. As such, the phrases "a bud and a leaf" or "a bud and two leaves" are used to describe the "leafiness" of a flush; they are also used interchangeably with pekoe and a leaf or pekoe and two leaves. Pekoe tea is a fine grade of tea which includes young tea leaves and buds. The tea once handled and brewed has "a rich forest-like scent with a hint of bitterness and a sweet finish."

ETYMOLOGY

The origin of the word "pekoe" is uncertain. One explanation is it is derived from the transliterated mispronunciation of the Amoy (Xiamen) dialect word for a Chinese tea known as "white down/hair". This is how "pekoe" is listed by Rev. Robert Morrison (1782–1834) in his Chinese dictionary (1819) as one of the seven sorts of black tea "commonly known by Europeans". This refers to the down-like white "hairs" on the leaf and also to the youngest leaf buds. Another hypothesis is that the term derives from the Chinese báihuâ "white flower" and refers to the bud content of pekoe tea

Sir Thomas Lipton, the 19th-century British tea magnate, is widely credited with popularizing, if not inventing, the term "orange pekoe", which seems to have no Chinese precedents, for Western markets. The "orange" in orange pekoe is sometimes mistaken to mean the tea has been flavoured with orange, orange oils, or is otherwise associated with oranges. However, the word "orange" is unrelated to the tea's flavor. There are two explanations for its meaning, though neither is definitive:

1. The Dutch House of Orange-Nassau, now the royal family, was already the most respected aristocratic family in the days of the Dutch Republic, and came to control the de facto head of state position of Stadtholder of Holland and Zealand. The Dutch East India Company performed a central role in bringing tea to Europe and may have marketed the tea as "orange" to suggest association with the House of Orange.
2. Color: The copper colour of a high-quality, oxidized leaf before drying, or the final bright orange colour of the dried pekoes in the finished tea may be related to the name. These usually consist of one leaf bud and two leaves covered in fine, downy hair. The orange colour is produced when the tea is fully oxidized.

FANNINGS

Fannings are small pieces of tea that are left over after higher grades of tea are gathered to be sold. Traditionally these were treated as the rejects of the manufacturing process in making high quality leaf tea like the orange pekoe. Fannings with extremely small particles are sometimes called dusts. Fannings and dusts are considered the lowest grades of tea, separated from broken-leaf teas which have larger pieces of the leaves. However, the fannings of expensive teas can still be more expensive and more flavourful than whole leaves of cheaper teas.

This traditionally low quality tea has however experienced a huge demand in the developing world in the last century as the practice of tea drinking became popular.

Tea stalls in India and the South Asian sub-continent, and Africa prefer dust tea because it is cheap and also produces a very strong brew - consequently more cups are obtained per measure of tea dust. Because of the small size of the particles, a tea infuser is typically used to brew fannings. Fannings are also typically used in most tea bags, although some companies sell tea bags containing whole-leaf tea. Some exporters focus primarily on broken leaf teas, fannings, and dusts.

GRADE TERMINOLOGY

- Choppy contains many leaves of various sizes.
- Fannings: are small particles of tea leaves used almost exclusively in tea bags.
- Flowery: consists of large leaves, typically plucked in the second or third flush with an abundance of tips.
- Golden Flowery: includes very young tips or buds (usually golden in colour) that were picked early in the season.
- Tippy: includes an abundance of tips.

WHOLE LEAF GRADES

The grades for whole leaf orthodox black tea are: Ceylon orange pekoe (OP) grades'

- OP1—slightly delicate, long, wiry leaf with the light liquor
- OPA—bold, long leaf tea which ranges from tightly wound to almost open
- OP—main grade, in the middle between OP1 and OPA, can consist of long wiry leaf without tips
- OP Superior—primarily from Indonesia, similar to OP
- Flowery OP—high-quality tea with a long leaf and few tips, considered the second grade in Assam, Dooars, and Bangladesh teas, but the first grade in China

- F OP1—as above, but with only the highest quality leaves in the FOP classification
- Golden Flowery OP1—higher proportion of tip than FOP top grade in Milima and Marinyn regions, uncommon in Assam and Darjeeling
- Tippy Golden F OP—the highest proportion of tip, main grade in Darjeeling and Assam
- TGF OP1—as above, but with only the highest quality leaves in the TGFOP classification
- Finest TGF OP—highest quality grade (Note: "Special" is occasionally substituted for "Finest", with a number 1 at the end to indicate the very finest), often hand processed and produced at only the best plantations, roughly one quarter tips
- SFTGFOP(1)—sometimes used to indicate the very finest

A joke among tea aficionados is that "FTGFOP" stands for "Far Too Good For Ordinary People".

BROKEN LEAF GRADES

- BT—Broken Tea: Usually a black, open, fleshy leaf that is very bulky. Classification used in Sumatra, Ceylon(Sri Lanka), and some parts of Southern India.
- BP—Broken Pekoe: Most common broken pekoe grade. From Indonesia, Ceylon(Sri Lanka), Assam and Southern India.
- BPS—Broken Pekoe Souchong: Term for broken pekoe in Assam and Darjeeling.
- FP—Flowery Pekoe: High-quality pekoe. Usually coarser with a fleshier, broken leaf. Produced in Ceylon(Sri Lanka) and Southern India, as well as in some parts of Kenya.
- BOP—Broken Orange Pekoe: Main broken grade. Prevalent in Assam, Ceylon(Sri Lanka), Southern India, Java, and China.
- F BOP—Flowery Broken Orange Pekoe: Coarser and broken with some tips. From Assam, Ceylon(Sri Lanka), Indonesia, China, and Bangladesh. In South America coarser, black broken.
- F BOP F—Finest Broken Orange Pekoe Flowery: The finest broken orange pekoe. Higher proportion of tips. Mainly from Ceylon's "low districts".
- G BOP—Golden Broken Orange Pekoe: Second grade tea with uneven leaves and few tips.
- GF BOP1—Golden Flowery Broken Orange Pekoe 1: As above, but with only the highest quality leaves in the GFBOP classification.
- TGF BOP1—Tippy Golden Flowery Broken Orange Pekoe 1: High-quality leaves with a high proportion of tips. Finest broken First Grade Leaves in Darjeeling and some parts of Assam.

FANNINGS GRADES

- PF—Pekoe Fannings
- OF—Orange Fannings: From Northern India and some parts of Africa and South America.
- FOF—Flowery Orange Fannings: Common in Assam, Dooars, and Bangladesh. Some leaf sizes come close to the smaller broken grades.
- GFOF—Golden Flowery Orange Fannings: Finest grade in Darjeeling for tea bag production.
- TGFOF—Tippy Golden Flowery Orange Fannings.
- BOPF—Broken Orange Pekoe Fannings: Main grade in Sri Lanka, Indonesia, Southern India, Kenya, Mozambique, Bangladesh, and China. Black-leaf tea with few added ingredients, uniform particle size, and no tips.

DUST GRADES

- D1—Dust 1: From Sri Lanka, Indonesia, China, Africa, South America, and Southern India.
- PD—Pekoe Dust
- PD1—Pekoe Dust 1: Mainly produced in India.

OTHER TERMS

- Musc.—Muscatel
- Cl.—Clonal
- Ch.—China varietal
- Qu.—Queen jat
- FBOPF Ex. Spl.—Flowery Broken Orange Pekoe Fannings (Extra Special)
- FP—(Flowery Pekoe)
- PS—Pekoe Souchong
- S—Souchong
- BOF—Broken Orange Fannings
- BPF—Broken Pekoe Fannings
- RD—Pekoe Dust/Red Dust
- FD—Fine Dust
- GD—Golden Dust
- SRD—Super Red Dust
- SFD—Super Fine Dust
- BMF—Broken Mixed Fannings

ISO 3103

ISO 3103 is a standard published by the International Organization for Standardization (commonly referred to as ISO), specifying a standardized method for brewing tea, possibly sampled by the standardized methods described in ISO 1839. It was originally laid down in 1980 as BS 6008:1980 by the British Standards Institution. It was produced by ISO Technical Committee 34 (Food products), Sub-Committee 8 (Tea).

The abstract states the following:

- The method consists in extracting of soluble substances in dried tea leaf, containing in a porcelain or earthenware pot, by means of freshly boiling water, pouring of the liquor into a white porcelain or earthenware bowl, examination of the organoleptic properties of the infused leaf, and of the liquor with or without milk, or both. This standard is not meant to define the proper method for brewing tea, but rather how to document tea brewing procedure so sensory comparisons can be made. An example of such test is a taste-test to establish which blend of teas to choose for a particular brand or basic label in order to maintain a consistent tasting brewed drink from harvest to harvest. The work was the winner of the parodic Ig Nobel Prize for Literature in 1999.

To maintain consistent results, the following are recommendations given by the standard:

- The pot should be white porcelain or glazed earthenware and have a partly serrated edge. It should have a lid that fits loosely inside the pot.
- If a large pot is used, it should hold a maximum of 310 ml (±8 ml) and must weigh 200 g (±10 g).
- If a small pot is used, it should hold a maximum of 150 ml (±4 ml) and must weigh 118 g (±10 g).
- 2 grams of tea (measured to ±2% accuracy) per 100 ml boiling water is placed into the pot.
- Freshly boiling water is poured into the pot to within 4–6 mm of the brim. Allow 20 seconds for water to cool.
- The water should be similar to the drinking water where the tea will be consumed
- Brewing time is six minutes.
- The brewed tea is then poured into a white porcelain or glazed earthenware bowl.
- If a large bowl is used, it must have a capacity of 380 ml and weigh 200 g (±20 g)
- If a small bowl is used, it must have a capacity of 200 ml and weigh 105 g (±20 g)
- If the test involves milk, then it is added before pouring the infused tea.
- Milk added after the pouring of tea is best tasted when the liquid is between 65 - 80 °C.
- 5 ml of milk for the large bowl, or 2.5 ml for the small bowl, is used.

CONTROVERSY

The protocol has been criticized for omitting any mention of prewarming the pot.

COMPETING STANDARDS

In 2003, the Royal Society of Chemistry published a news release entitled "How to make a Perfect Cup of Tea."

TEA LEAF GRADE AND CAFFEINE CONTENT

Contrary to popular belief, black teas don't always have more caffeine than green, white, or oolong teas. In fact, certain green and white teas can have significantly more caffeine than black teas. Caffeine content has more to do with the part of the tea tree that it comes from. The topmost bud contains the most caffeine. The topmost leaves contain the next highest caffeine levels. The larger, lower leaves contain the least caffeine. Typically, higher end teas will have more tea buds. A bolder cup of tea will use lower leaves on the tea tree. Surprisingly, a stong, smoky Lapsang Souchong will have a low level of caffeine compared to a tippy white tea such as Silver Needle. Of course, there are other factors that affect the level of caffeine in tea, such as steeping time and temperature. Stay tuned for our next article on all the factors that affect caffeine level in tea.

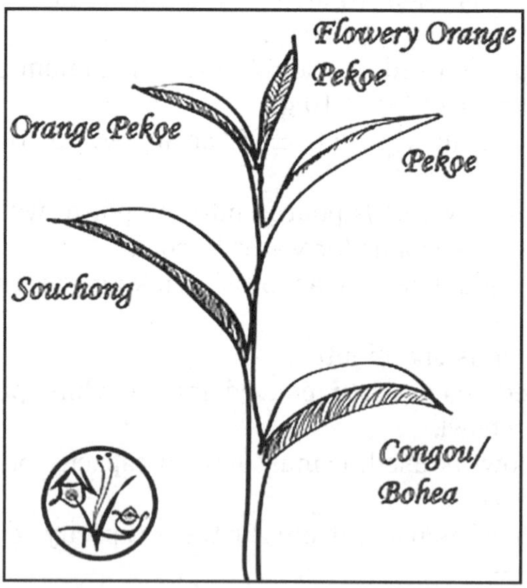

CAFFEINE LEVEL BY PART OF TEA TREE

Each part of the camellia sinensis plant (tea tree) has different caffeine levels.

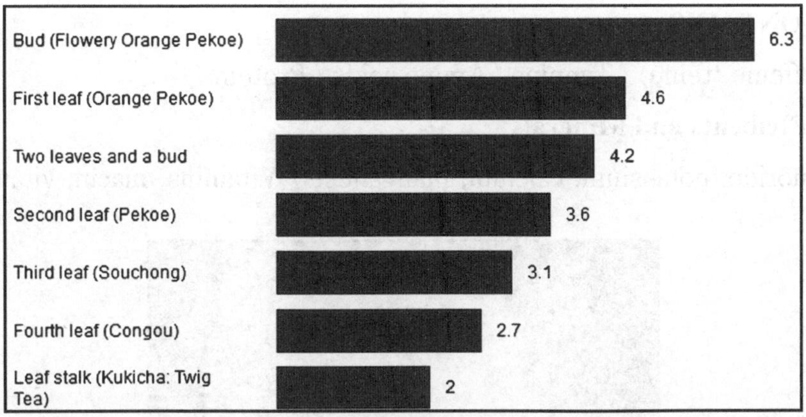

TEA LEAF GRADES & PRODUCTION METHODS

THE PLANT

The tea plant (camellia sinensis) is a species of tree related to the camellia. Its flowers are yellowy-white and its fruits small and hard-shelled, similar to a hazelnut. The evergreen leaves are leathery, dark and slightly serrated. Given minimum annual temperatures of 18° C, moderate and infrequent frosts, a uniform annual precipitation of 1600 l and a good balance of sunshine, a tea plant can easily grow to become 100 years old. Wild tea plants are indeed reputed to reach an age of up to 1,700 years.

Two Original Tea Plants are Known Today

Thea sinensis (or Chinese tea): A shrub-like plant which reaches a maximum height of 3-4 m and can even survive frosts. Thea assamica (or Assam tea): A substantial tree reaching a height of 15-20 m which grows exclusively in the tropics. The constant crossing of these two original plants forms the basis of all the tea cultures in the world today.

ACTIVE INGREDIENTS IN TEA

The cheapest drink worldwide after water is one of the most valuable in terms of its chemical composition: approx. 32 % of its ingredients pass into the infusion.

TEA CONTAINS

Caffeine (teine) / Tannins / Amino acids / Proteins

Trace Elements and Minerals

Fluoride, potassium, calcium, manganese / Vitamins: niacin, vitamin B1 and B2

"Tea both stimulates and calms": Tea owes its stimulatory effect to its caffeine (teine) content: It does not act on the circulation via the heart, however, but directly on the brain and central nervous system, as it is bonded to the tannins and is not released until it reaches the intestine. This explains the demonstrable capacity of tea to increase concentration and responsiveness. Cultivation: Tea is propagated by taking cuttings from parent plants.

The tea plant is kept in the vegetative phase by regular pruning to prevent flowering and fruit formation. This also makes it easier for the tea pickers to gather the 2 uppermost leaves and the newest bud (only these are relevant for the tea harvest). Most picking is still done by hand in order to preserve the quality of the harvest. Some countries have developed mechanical picking methods, however, which greatly simplify production processes. Tea is cultivated in large plantations/estates such as those of the Indian district of Darjeeling, e.g. on the 2,000m high slopes of the Himalayas. Much lower lying is the Assam region of Northern India, the biggest continuous tea cultivation area of the world, situated on either side of the Brahmaputra.

On the island of Sri Lanka (formerly Ceylon), a bitter, aromatic tea is produced in the three cultivation areas of Dimbula, Nuwara Eliya and Uva which has become famous throughout the world. The Chinese produce a distinctly smoky, mild tea in various provinces such as Yunnan and Zhejiang. The Chinese are known for their tea specialties, such as tea bricks or tea roses. Japan produces exclusively green tea, most of which is consumed by the Japanese themselves. Other tea-producing countries are Africa, Indonesia, Taiwan and Argentina as well as Thailand, Russia and Turkey, though these are relatively insignificant in terms of production volume.

THE ART OF GRADING TEA

Tea grading is one of the most complex topics in the industry. Specialty imbibers are routinely confused about the topic. Randy Altman explains this intricate classification process.

There is no universal standardized grading system employed throughout the global business, and even those systems that do exist are often more art than science. Companies also occasionally fool around with grades, giving an exalted grade to an average tea. However, perhaps the most well-known grading system in America is actually based not on quality, but on leaf size. This system revolves around the classification of Pekoe, for which the letter "P" is the designation.

Pekoe simply means a whole leaf was used in the manufacturing process. There is in the "P" grade no meaning of quality. The system that uses Pekoe as its base refers mostly to tea from India and Ceylon, almost always black tea. The next grade, and they do come in steps or a series, is "Orange" making for "OP." This is a slightly larger leaf that may also be thinner. The word "orange" certainly adds to the confusion, as the color bears no relationship to the use of the term. Many tea books or experts in lectures will say the term derives from the Dutch royal family Orange Nassau, but actually the origin is obscure and this is just a convenient guess. Perhaps further confusing to Americans and Europeans is the grading system under discussion mainly reads from right to left, the opposite of everyday text reading, which is left to right. The next designation starts to enter the quality realm. The subsequent appellation is "flowery" and hence we have "FOP." And yes, more room for confusion among the retail consumer, since this "flowery" does not refer to any flower, rather the unopened bud of a tea leaf. All tea plucked by hand should at a minimum be "FOP," as a generalization. To review, FOP stands for Flowery Orange Pekoe.

The next grade is "G" for "golden," an FOP in which certain leaves have a golden tip, indicating higher quality. Thus we have GFOP. The next category is tippy and all tips should be golden, according to authoritative experts. The designation is TGFOP. Here we have the beginning of more frequent over-exaggeration of grading by some companies, as many teas sold as TGFOP are not all golden tipped. Getting into the truly super-premium specialty tea realm, we move into "finest" which yields the designation "FTGFOP." This adds another potential element of confusion, as we now have two "F's" in the one grade, but they stand for different words. We are not yet done with this classification system. The final word to add to the left of the starting word Pekoe is "special" which like all the other words goes by its first letter, "S" resulting in "SFTGFOP."

Tea can be yet more glorious, and here the exalted grades return to the beginning "P" and are placed to the right of the "P." Probably the most common designation is the number 1, for "SFTGFOP-1," and I have certainly gotten samples of this grade from Darjeeling that merit the label. Some companies are not satisfied with the humble number, and add terms like "extra special" or "extra fancy" for a really long grade of "SFTGFOP-Extra Fancy." I have no idea what percentage of specialty tea so graded actually deserves such lofty titles. If a company over-grades, the specialty imbiber can vote with their feet and change shops, catalogues or websites. This grading system is much more complex than the above discussion implies, primarily because tea leaves are so often chopped up into pieces. The size below pekoe is "broken pekoe" or "BP" and yes, you can have for example, a golden flowery broken orange pekoe, a "GFBOP." The next lower size is fannings, and finally there is dust. Fannings and dust are almost exclusively for tea-bags of the regular size tea-bag. Here is one of the most commonly held myths about tea grades. I hear this one all time. The myth is that dust and fannings are somehow bad tea. The best tea will yield some percentage of dust and fannings. These fannings and dust will make a good brew, not the greatest, but a good brew. Now, freshness must be factored in more with dust and fannings than with whole-leaf teas, because the tiny particles have a worse surface-to-air ratio and lose freshness faster.

The exalted grades of these black teas are more art than science. As with any super-premium tea, these tend to come from named estates. The retail consumer needs to be aware that a purchase one year that is perfectly satisfying does not signal the estate will automatically produce such a fine tea in following years. Many variables can intervene over time, something as simple as not enough rain. Some variables associated with quality tea have nothing to do with grading, such as organic, kosher, and FairTrade. This is yet another potentially confusing aspect of grading to the paying imbiber.

This Pekoe-based grading system is by no means followed universally even in India and especially Ceylon. In Ceylon, one often encounters a simple three-tier system of low-grown, medium-grown, and high-grown. I have had very nice low-grown Ceylon teas. Ceylon also tends to alter the above "pekoe" system more often than India. For example, Sri Lanka Tea Board mentions a "Dust 1" which they describe as "fine granular particles that bring out the optimum strength and body, ideal for commercial brewing." Sri Lanka Tea Board even mentions an "FBOPF1" (Flowery Broken Orange Pekoe Fannings 1). This category they describe as a "typical low country semi-leaf tippy tea with a full-bodied sweet brew."

This discussion over-simplifies this one grading system, but covers the basics. I will venture into the weird rule of most connoisseurs that "pekoe" be pronounced as "peck-oe" and such pronunciation is taken as sign of a true aficionado. I have studied ancient language up to the post-graduate level, although not Chinese, where the word originates. I own two giant dictionaries with very long distinguished names, too long to mention, and both show only one pronunciation for the tea word "pekoe" and the proper usage is as the word looks like it should be verbalized, "pee-koe." Tea books often go out of their way to mention to show your tea intelligence by saying "peck-oe," but I swear I remain counter-trendy as per my beloved dictionaries. Mention should be made of a few differences between Darjeeling and South India. Darjeeling follows the "pekoe" grading system rather closely, more closely than most other regions. You might see a "CL" suffix, which stands for "clonal" and again this is not necessarily a quality plus, but in my experience it usually means the owner is paying good attention to the tea plant. The main complicating factor with Darjeeling is that it has distinct growing seasons. "First Flush" are springtime teas, and the word most often associated with these young leaves is muscatel. The next round is "Second Flush" and is more mature, which is a plus if you prefer this. Sometimes a third round is mentioned, called "In-Between Darjeeling" and is harvested in April and May. Finally, there is a last round, called "Autumnal," known for its large leaves.

South India follows the Pekoe grading system less closely, largely because it tends to use the CTC (cut-tear-curl) method of harvesting, as opposed to traditional hand-plucking. The largest plantation company in South India, Harrisons Malayalam, recently gave me a wide range of fresh samples. One was graded "CTC Spc" with the suffix meaning "special." Similarly, another was "fannings supreme." Closer to the Pekoe system was samples marked, "BOP Supreme" and "TFBOP" and one a bit different, "Spc PF." Several others had traditional pekoe gradations. Finally, a "Dust spc."

A completely different system of grading is used for green tea. Actually, multiple systems are used for green tea, and again we must over-simplify, but will cover the basics. Here, grades closely refer to quality, especially when employed by Japan. The following is the Japanese system. The lowest grade is Kukicha, and is in fact made from the twigs and stems of the tea plant, and historically given to children and seniors, and is not usually sold in America. The next higher grade is Bancha, and is still rather low in quality and not sold, at least should not be sold, by specialty tea companies. The third grade is Sencha, and now we have a very wide range. The specialty tea buyer needs to be careful if paying a lot of money for a product labeled "sencha" as this may or may not be appropriate. Japanese sencha can be superb, and now China and Viet Nam make sencha, with more mixed results.

The next higher grade, a truly wonderful experience, is Gyokuro, which is usually translated as Pearl Dew. This tea is very labor-intensive, and is shade-grown for part of the year. This tea is expensive, but worth it. The quantity is low, that is, a small output, picked but once a year, with small, tender, extremely fragrant leaves.

The fifth grade, the highest, is matcha (often spelled "mattcha"). This is simply the best gyokuro pulverized into a fine powder and used during the famous Japanese Tea Ceremony. Here, the tea is not steeped, but whisked directly into the water, making a frothy brew of delight.

China is a huge nation, unlike Japan, and has hundreds and hundreds names for tea, without a real system. These names reference estates, but the translations of the names often refer to the appearance of the leaf, not a location or owner. The closest thing to a grade of tea in China is "Hyson" or "Young Hyson," and all sorts of teas are thusly labeled. "Hyson" was not originally even a category of Chinese tea, but the name has caught on and become trendy. If you are dealing with a good specialty tea company and it lists a Hyson, odds are the Hyson will also be satisfactory. As for superlatives used as adjectives, like the "extra fancy" for black tea, the term "Imperial" is sometimes used for Chinese.

WHY ARE TEA GRADES IMPORTANT?

Sometimes we come across tins or packets of tea that are labelled with different letters along with their name. These acronyms are "grading terms" added to the name by the producer to give extra information about that tea. These terms don´t refer to sensory aspects of the tea –they don´t describe the body, texture or aroma of a tea- but give information about the appearance of the leaf that may be helpful when buying teas from a catalogue. These grades are only applied to black teas from certain countries. They vary from country to country and even my best students can get confused, but if we learn how to read the grading names and their acronyms, shopping for tea can become a little easier.

There is no one universal grading system. For example in China, teas are usually graded by number, with 'first' being the highest grade, down to 7, 8 or 9, based on leaf style and shape and how carefully the manufacturing process has been carried out. In Taiwan and Japan, grading terms range downwards from, for example, Extra Choicest, Choicest, Choice, Finest, Fine, Good Medium, Medium, Good Common, Common, Nubs, to Dust and Fannings. The most widely-used grading terms are the ones applied to teas produced in other places, like India, Sri Lanka, Africa, Argentina, Indonesia, Malaysia, and Europe.

For broken leaves, the letter "B" is added to the name as in BOP (Broken Orange Pekoe), FBOP, GBOP, TGBOP, etc. For smaller grades used in teabags, fannings and dust grades are used, for example: OF, OPF, FBOPF, FD, GD. It is important to mention that tea grades are designated by the producer, and sometimes can vary from one to another. So asking for samples before buying is a way of making sure you are buying the grade you are expecting. So, going back to our initial question, why do you think tea grades are important? They are important because learning how to read these acronyms and their meanings they provide a lot of information about the tea itself, very useful when it comes to choosing a tea profile.

We know small pieces of leaf give stronger liquor than whole leaf teas. So if you are looking for a black tea to be used in a breakfast style blend, then you should go for a broken tea (look for the "B" in the grade). On the contrary, if you look for a mellow, soft and sweet black tea, one with high quantity of buds will provide such an experience, so pick the ones with an F or G letter. F stands for flowery –meaning that the tea has buds- and G stands for Golden -which denote a higher proportion of buds on the tea.

Leaf grades help to understand how the leaves look like and what to expect from them regarding strength, aroma, sweetness...All the information we can gather about the teas will be very useful when making a purchase or when starting a business. Very often tea purchases are made from a catalogue, by studying a long list of teas sent by the seller. The abbreviations, designations and grades previously discussed will help us get an idea of the appearance and style of the tea, but remember that to make a good purchase is advisable to request samples and cup each tea before buying.

SORTING AND GRADING OF TEA

Sorting is a mechanical process. In the early days of the industry, it was believed that sorting did not require the same degree of management as other operations of tea manufacture. But, it has been found that in order to get good prices at the auction, proper sorting is essential for all grades of tea in general, and broken grades in particular.One of the objectives of sorting is to classify tea according to the size of the particles. In the past, the object of sorting was merely to remove foreign mater and big leaves.

This process was carried out by hand, with the only implement used being a small pincer used to pull out long stalks. The process of grading was known as sifting at that time. Today, no distinct division can be made between sorting, cleaning and grading since the processes are carried out simultaneously and procedures are common to all. All processes are based on the four important characteristics of made tea.

These include:

- Size,
- Shape,
- Specific gravity, and
- Surface characteristics of the particles.

MODERN METHODS OF SORTING

Modern methods of sorting were introduced along with mechanisation in the 1870s. Mechanical sorters were developed to utilize power from steam engines and other sources. In 1880, George Reid introduced a mechanical breaker into the sorting operation to reduce the size of leaves. In one stroke, a very tedious process was greatly simplified.

More modern sorting machines incorporate different sieves with meshes of size 8,10,12,16,24,30 or greater. Tea passed over a number 8 mesh is cut and sifted again, and so on. As the tea passes through the various meshes, it is classified into grades such as whole leaves, broken leaves, fanning, and dusts.

In modern factories a Myddleton machine is used at the first stage of the operation for orthodox teas. For CTC manufacture, an electrostatic stalk extraction unit is used first. Usually, in orthodox manufacture, the spillover from the Myddleton is passed through a ghugi or rotary hexagonal sifter; from the trays, the leaf is passed into the Arnott and Macintosh sorters. But in CTC and leggcut manufacture, the ghugi is not used.

Thereafter, the process incorporates the Britannia Tea Sorter (which works very well in duel manufacture), the Macintosh sorter and the wind tunnel. A number of hand sievings are also essential to produce some quality grades. In wind tunnels and waterfalls, fibre and dust grades are separated according to the specific gravity of the particle. Tea breakers and cutters are extensively used in modern sorting operations to reduce particle sizes for subsequent classification. In orthodox manufacture, the first classification of tea takes place at the time of sifting. BOP, BOPF and OP grades are produced from the first and second fine teas. The coarse leaf that is first broken or cut goes into the production of BP, PF and souchong grades. In CTC and leggcut manufacture, sorting is based wholly on the size of the leaf particles.

Precaution to be Taken while Sorting

Sorting should not be carried out directly on hot particles discharged from the dryers; this can destroy the bloom of the made tea. After drying, the particles should be cooled on the floor or on a wooden platform. The tea should not be bulked hot. At the same time, care should be taken to see that the leaf does not cool to the point where it can pick up atmospheric moisture.

Grades of Tea

Broadly speaking, there are four grads of tea. These whole leaf, broken leaf, fannings and dust. Each of these grads again have various sub-grades. Of the four, only the latter three can be produced via CTC and leggcut manufacture.

Orthodox Whole Leaf Grades

Golden Flowery Orange Pekoe	GFOP
Flowery Orange Pekoe	FOP
Orange Pekoe	OP
Pekoe	P
Pekoe Souchong	PS

Orthodox Broken Grades

Golden Flowery Broken Orange Pekoe	GEBOP
Flowery Broken Orange Pekoe	FBOP
Broken Orange Pekoe	BOP
Broken Pekoe	BP
Broken Pekoe Souchong	BPS

Orthodox Fanning

Flowery Orange Fanning	FOF
Orange Fanning	OF
Pekoe Faning	PF
Fanning	F
Broken Orange Pekoe Fanning	BOPF
Broken Pekoe Fanning	BPF
Pekoe Dust	PD
Red Dust	RD

CTC Grades

CTC Broken

Broken Pekoe one	BP1
Broken Pekoe two	BP2
Broken Orange Pekoe	BOP
Broken Pekoe Souchong	BPS

Fannings

CTC Pekoe Fannings one	PF1
CTC Pekoe Fannings two	PF2

CTC Dust

Pekoe Dust	PD
Dust	D
Churamoni Dust	CD

Leggcut Grades

Leggcut Fannings

Orange Fannings	OF
Pekoe Fannings	PF

Leggcut Dust

Dust one	D1
Dust two	D2
Churamoni dust	CD

PERCENTAGE OF VARIOUS GRADES

The types and percentages of various grades produced in a tea factory depend upon market demand. Also, but to a lesser extent, plucking standards and the quality of leaf play a role. In orthodox tea manufacture, the following percentages of different grades are normally produced:

Whole leaf	15%
Broken	60%
Fannings	18%
Dust	4%
Residue	3%

From the 15 % whole leaf grades, the following three sub-grads are extracted.

FBOP	2%
OP	7%
P	6%

Similarly, the broken, fanning and dust grades are processed further into:

FBOP	15%
BOP	36%
OF	10%
BP	7%
BPS	2%
PF	8%
Dust	4%

It should be noted here that failure to produce whole leaf grades will result in reduced liquor colour and strength in broken and fanning grades. In CTC tea, primary broken and fanning grades constitute about 20 and 50%, respectively.

Secondary siftings of the same two grades yield about 9% each. Dust accounts for about 18% of the total produt, leaving a residue of 3%. In comparison, leggcut manufacture yields fewer grades, making it simpler than orthodox and CTC manufacture. In leggcut manufacture, only fanning and dust grads are produced, in the following percentages:

Fannings	80%
Dust	19.25%
Residue	0.75%

In green tea manufacture, various grades are produced in addition to the common fanning and dusts. These include:

Hyson No.1	90%
Jwamkay	2%
Fanning	2%
Dust	5%

Stalk Extraction

The extraction of stalk from made tea is one of the major problems in the sorting process. Stalk-free tea can be produced through fine plucking, light rolling, and avoiding drastic treatment to the leaf during processing. Unfortunately, such procedures are not always economical. Red stalk and fibre often appears and their removal becomes necessary. At present, the Myddleton is the machine most commonly used for stalk extraction. In this machine, there are two trays with bosses on the surface. During processing, the particles are hopped through these bosses and are thus separated. Java tunnels and waterfalls, which operate on the varying specific gravity of tea particles, are also useful in stalk removal. There is also the more recently developed Andrew Breaker, with its flute rollers which break leaf up but leave stalks intact. Japanese tea technologists have also invented a system based on the variation of moisture levels between leaf and stalk particles. Though the machine works well in principle, its output is low to cope with the needs of a modern factory. This, however, is true of most stalk extraction systems. The presence of stalk and fibre is controlled at best, not eliminated. A fine balance needs to be maintained between the level of control, efficiency and market price.

Greying of Tea

In the buyers' market, tea is valued for colour, strength quality and briskness. In order to obtain fine tea with all the above desirable characteristics, quality must be inherent in the leaf when it isfirst brought to the factory. No skill can produce high class tea out of poor quality leaf. By the same token, poor manufacture can easily turn good leaf into bad tea. One defect, arising from flawed manufacture, is greyness or the loss of bloom. This accounts for significant losses for the tea industry. But, before dealing with details of the subject, let us understand the terms 'bloom' and greyness'.

Bloom

The bloom sought in tea leaf is achieved through the development of a varnish-like film on its outside surface. Failure to produce proper bloom leads to dullness in made tea. Teh bloom can be damaged through faulty sorting.

Greyness

Grey is perhaps the most undesirable colour in dry leaf. greyness is caused when the gummy or varnished surface of the leaf, which plays an important part in its liquoring properties, is abraded or rubbed off. greyness is particularly likely to occur during sorting and may be caused by one of the following factors.

- Tea particles get caught in sagging meshes in the sorting machine and are abraded through the movement of the trays.
- Blunt knives in the cutter and some older types of machine break, rather than cut, the tea. This not only produces excess dust, but also causes greying. Fast running cutters too tend to cause greyness.
- Excessive handling of the product also creates greyness.
- The tea readied for sorting should have a moisture content of between 3 and 5%. Moisture above 5% causes greyness.
- Over-sorting quickly causes greyness because of the continuous rubbing and striking action of the particles against the trays. A leaf's varnish is easily removed in this manner.
- Excessive rolling may also cause greyness.
- Tea leaf fed into a sorter from an incorrectly placed hopper is subject to striking action that also removes the outer varnish.

SORTING PROCEDURE

Every tea estate has its own method of sorting. This is because each company has to develop its own criteria, depending on the market it services. But there are some common procedures that most gardens would follow for CTC and orthodox manufacture. In CTC manufacture, tea is initially passed through an electrostatic fibre extractor. Once the fibre and stalks are removed, the tea is fed to the Myddleton machine. From here, the tea passes to one or two successive Arnotts, a Macintosh machine for final grading, and then it is cleaned in a Java tunnel or waterfalls. The spillover from the Myddleton is passed through the Andrew Breaker and recycled for grading. The first step in orthodox sorting also involves the involves the Myddleton. Teh output from its the two trays is passed into the Macintosh, where the teais graed through various meshes. The spillover from the Myddleton is passed to the Ghugi, or rotary hexagonal sifter, where the TGFOP, GFOP and FOP grads are extracted. But the sequence of grading varies widely from estate to estate. For example, in some estates, fine tea from dryers is fed directly to a Ghugi fitted with number 14,12 and 10 meshes and sorted in the following manner through hand sieving.

Sorting of Leggcut Tea

Leggcut tea is easier to sort than orthodox and CTC teas. The bulk of the work is done in the Myddleton and Macintosh machines, where the output from the former passes to the latter to produce fanning and dust grades.

SORTING MACHINES

Myddleton

The Myddletontea stalk extractor is a common machine in all types of tea manufacture. It is used widely in almost all tea growing areas as an initial sorting machine. It performs the operations of grading and making tea free of stalks.

Important Parts

- Trays: There are two trays in the Myddleton, the top and bottom trays. The top tray is 9' x 4' and the bottom tray is 8'5 inch x 3'10 inch. The trays are made of thick aluminum sheet, with lines of small bosses, each with a perforation at the top. The trays are easily removable and sieves of different specifications can be fitted into the frames to suit the tea being treated. One of the Myddleton's two sieves is normally has 3/16 inch perforations and 1/8 inch high bosses. Two additional trays are also available; one with 7/32 inch perforations and 1/8 inch high bosses, and the other with 17/64 inch perforations and _inch high bosses.
- Pulley: There is a 10 inch pulley with a 3 inch face, suitable for a 2.5 inch drive. This can be attached directly to a motor.
- Crank shaft: The crankshaft imparts reciprocating motion to the machine via connecting rods. The crankshaft is made of very strong steel bars.
- Grafton Phillips Attachment: The Grafton Phillips Attachment is used for removing dust and fluff from the tea. It has two superimposed trays, fitted with brass wire mesh number 16 on the top and number 32 at the bottom. The tea discharged from the top tray of the extractor is passed to the top tray of this attachment. An oscillating motion causes the tea to be thoroughly sieved as it passes over the gauge. The finer parts of the tea and the dust fall onto the lower tray. On this tray, a similar operation takes place - the cleaned fine tea passes over the end of the tray while the dust falls through into a collector underneath. The tea discharged over the ends of both trays is clean and free of dust and fluff.
- Working principle: An oscillating motion is imparted to the sieves wit the help of a crankshaft, which revolves at 200rpm. Leaf is fed to the Myddleton machine at spread thicknesses of approximately 1.3 to 1.9 inches. The tea being sorted in the Myddleton moves down a slopped tray between the bosses. This imparts a slight disturbance to the mass and causes the leaf to hop through the perforations. The stalk and fibre slide over the bosses pass over the end of the tray, and is thus separated. The Myddleton can also be used for grading

by treating fine and coarse teas separately. The following arrangement of trays would then the required.

- Power requirement: The Myddleton operates on 1 hp of power
- Capacity: The machine is capable of processing 300 kgs of leaf an hour.

Rotary Sifter

The rotary hexagonal sifter, or Ghugi, is one of the most important machines for orthodox operations. This machine has gained popularity for its efficient working. Normally, leaf sorted at the uppermost level of the Myddleton is fed to the Ghugi for selection into other grades.

Important Parts

- Screens: The important components of the Ghugi are three screens of different mesh combinations - such as 14,12 and 10 or 10,8 and 6 - forming a hexagonal body of 5 ft and 3 ft at two ends. The sifter is set on an incline, which can be adjusted according to requirements. The sifter is usually about 10 ft long.
- Working principle: The sifter can be run at 25 to 30 rpm with the help of a motor or a line shaft. Due to the rotation of the sifter at an inclined position, the tea moves from the feeding point to the sifter's discharge end. The tea passes through various separating zones in the sifter according to the size of the particles. Thus, the grading of tea is carried out.
- Power requirement : The Ghugi requires 3 hp of power
- Capacity: It processes about 1,000 kgs (or 2,200 lbs) of leaf an hour.

Electrostatic Fibre Extractor

The Electrostatic Fibre Extractor (EFE) is a relatively new machine used for stalk extraction after firing. In this machine, the principle of static electricity is used for extracting fibre and stalk. It can be used in continuous manufacture with the help of a conveyor system after the second firing. The machine is normally 10-12 ft long and 3-4 wide and is considered one of the best for extracting fibre in CTC manufacture.

Different Parts

- Main frame: The main frame of the EFE is made of mild steel, and is rigidly built. In some motor-driven models, cast iron wheels are attached to the frame so that the machine can be easily moved within the factory.
- Roller: The machine has seven PVC rollers, placed in two stages of four and three rollers, respectively, Each roller has a diameter of

about 4 inch, and is designed to rotate at speeds of 2,000 to 2,500 rpm. The rollers rotate smoothly on heavy duty bearings housed in plumber blocks.

- Motors: The motor used in this machine is of 1400/1500 rpm, 400/440 volts 3 phase 50 cycles.

Working Principle

This machine works on the principle of static electricity. There are seven PVC rollers arranged in two rows that run at speeds of 2,000-2,500 rpm. Tea fed into the machine via a conveyor system comes into contact with the PVC rollers, causing friction between the two.

As a result, static electricity is produced with the fibre and rollers developing opposite charges. This causes the particles and rollers to attract each other. But the fibre is turned over through centrifugal action and collected in aluminum collecting boxes fixed over the conveyors between rollers. From there it is transferred to other containers.

Power Requirement

The EFE uses 3 HP.

Capacity

The EFE process 400 to 450 kgs (880 to 990 lbs) of made ta an hours.

The Macintosh Tea Sorter

The Macintosh tea sorter has several outstanding features. It is widely used for orthodox. CTC and leggcut manufacture. The unique feature of this machine is that it makes uniform size grading possible. In today's tea market, this if of great importance.

Important parts :

- Crankshaft, Flywheel and Bearing: The crankshaft used in a Macintosh is known as a single throw crankshaft, and is made of a round mild steel bar. The crankshaft is fitted with a heavy-duty ball bearing which, in turn, carries a balance flywheel. This ensures vibration-free operation.
- Trays: The machine has five trays with seven discharge outlets for various grades of tea. The trays slide into the frame and are each securely fastened with four band-operated screws. All trays are interchangeable, and the mesh used depends on the type of tea to be sorted.

  The selection of trays too depends on th grades to be produced, but the following broad guidelines may be considered for sorting leaf.

Trey no.	Mesh no.	Grade discharged
1.	30 at closed end	No. 1 Brown Dust
	24 at open end	No. 2 Dust
2.	20	No. 3 Fanning
3.	12	No. 4 Broken Pekoe
4.	19	No. 5 Small Pekoe
5.	8	No. 6 Larger Pekoe
		No. 7 Pekoe Souchong

In the case of broken grade teas, the arrangement of trays can be made as follows

Trey no.	Mesh no.	Grade discharged
1.	30 at closed end	No. 1 Brown Dust
	24 at open end	No. 2 Dust
2.	20	No. 3 Small Fanning
3.	18	No. 4 Larger Fanning
4.	14	No. 5 Small Broken Pekoe
5.	10 to 12	No. 6 Larger Broken Pekoe
		No. 7 Broken Peko2 Souchong

- Working principle: Tea is fed to the Macintosh through a hoper attached at one end. The rate of feed is controlled by adjusting a mechanically agitated flap valve. A satisfactory rate of feed has to be ensured to suit the different types of tea being sorted in the machine. The gradient trays are adjustable and quickly interchangeable so that grading can be changes as desired. The machine's frame is connected by a connecting rod to a single crankshaft fitted with heavy duty ball bearings. The crankshaft rotates at 250 to 270 rpm and transmits a straight oscillatory motion to the trays with the help of spring legs. Due to this motion, tea particles pass through holes according to their size, and are thus classified into various grades within 8 to 10 seconds.
- Power required: The Macintosh sorter uses 2 hp of electric power.
- Capacity: The Macintosh sorter processes approximately 220 kgs of tea an hour.

Balanced Tea Sorter

The most attractive feature of the Balanced Tea Sorter is that it can be operated efficiently even while maintaining the best bloom of tea. It is considered most suitable for the dual type of manufacture. Various sieves with different mesh sizes sort tea particles into the required grades. However, the drawback of the Balanced Sorter is that its output is lower than that of other comparable sorting machines.

Important parts:

- Hopper: A hopper fitted at one end of the Balanced Sorter contains no wearing parts, ensuring even input of tea. A flap easily regulates the thickness of spread. In the lower part of the hopper, there is a rubber diaphragm which is subjected to the slight impulse of the beater bar during the upward movement of the trays. The required impact distance is very small, with not more than 1/16 inch being needed for CTC tea and little more for orthodox tea. The action of the beater bar can be adjusted easily while the machine is running. The impact should be kept to a minimum, as this will prolong the life of the diaphragm. Two handles on both sides of the hoper are used to control the shutter. Two small glands grip the spindle of the shutter-opening device just tightly enough to prevent self-movement, and thus eliminate the use of any clamping or locking devices. While the hopper can be fed through a conveyor; it can also be hand fed. Since the hopper is stationary in this machine, greying can be avoided.

- Oscillating links : Ideally, the links that impart Jigger type motion to the machine, are set at an angle of 15°. The links are fitted with 'silent block' bushes, which do not require any lubrication during operation.
- Bearings: The ball bearings attached to the machine are properly enclosed to ensure protection from dust and moisture for at least six months at a time.
- Trays: This machine comes equipped with five trays, with the bottom-most used for producing dust grades. Half this tray's length is fitted with number 40 mesh, and the rest with number 30 mesh. The other trays are normally fitted with number 8,10,12, 20, 30 and 40 meshes. Since a simple spring clip automatically holds the trays in position, no clamping device is required to remove or the refit trays. This can easily be done manually.
- Working principle: The Balanced Tea Sorter receives power from a motor or line shaft belt. A barrel with a cambered portion runs at speeds of 300 rpm. This cambered portion serves as a fixed pulley and imparts Jigger-type motion to the machine's fitted screens via links. The lower slope of the machine's feed hoper is made of a rubber diaphragm forms. As tea passes over it into the machine, it is subjected to the impulse of the oscillating trays' beater bar. This ensures a uniform feed into the machine.
- Power: The total power required for this sorting machine sis 0.25 HP. but, a 1 HP motor is used to over come the initial torque.
- Capacity: The Balanced Tea Sorter can process up to 200 kgs (440 lbs) of tea an hour.

Benton Density Sorter

The Benton Density Sorter is a small machine that works on the density variations of particles. A small fan that consumes only 30 Watts of electricity powers this machine. Four wheels fitted to the machine make it easy to move the machine to a desired location anywhere within a factory.

The Benton Density Sorter is used for the following operations:

- Separation of sand from dust tea.
- Separation of fibre and flaky leaf from other grades.
- Separation of large foreign matters from dust tea.
- Separation of stones and broken grades.

Important parts:

- Fan and Motor: The Benton Density Sorter is powered by a small fan driven by a totally enclosed 0.125 HP motor that consumes approximately 30 Watts.
- Feed shutter: A feed shutter is used to admit leaf into the machine; the shutter can also control the rate of feed. The rate of feed is controlled at thickness of 0.25 to 0.5 inch.
- Chute: The chute is a polished aluminum sheet on which the leaf slides into the machine. It can set at various angles to suit the needs of different particles.
- Working principle: As soon as leaf is fed into the hopper, it is admitted to the chute through the feed shutter at 0.25 to 0.5 inch thickness. The chute is set at such an angle that the leaf accelerates rapidly down its surface in such a manner that individual particles separate out.

  At the point where the chute terminates in the gap, the leaf mass is subjected to a current of air supplied from beneath. The heavier particles are least affected by the current of air and pass into the side delivery number 1.Lighter particles float across the gap on the cushion of air and pass into side delivery number 2. Similarly, the lightest particles are carried upward for a short distance and pass into the end delivery.The chute can be angled from 30° for heave grades to 70° for powdery grades. This change is possible merely by altering the position of a clamp on its rack. The angel of the fan duct can not be changed relative to the angle of the chute; but there is a provision for altering the angle of the fan duct separately to optimise the airflow through the gap.

  Normally, the machine is operated with the fan duct in the lowest position. The volume of air used in this machine is very small, but that can be adjusted through orifice plates of varied diameter. Orifice plates of 9, 6 or 4 inch diameters can be used in conjunction with the two available fan speeds.
- Capacity: The Benton Density Sorter can process up to 10 kgs a minute.
- Power required: The machine runs on a small 0.125 HP motor.

Circular Vibro Screen

The circular Vibro Screen was introduced into tea industry for the classification of various grades of tea in an efficient and economical way. This machine is designed to work in a continuous manufacturing situation. Essentially, it comprises a circular gyratory screen that is subjected to three-component - horizontal, vertical and tangential - multiplane mechanical vibrations. The vibration is created by a motor with a double extended shaft fitted at both ends with eccentric weights. Both motor shaft and weights are mounted on the screen assembly. The drive assembly, in turn, is mounted on a circular base fitted with rugged springs. The springs not only allow the unit to vibrate freely, but also prevent the vibration from reaching ground where the machine is installed.

- Working principle: A three-component multiplane mechanical vibration is induced by a motor mounted on a vertical axis. The motor has double extended shafts fitted at both ends with eccentric weights. It is suspended from the screen assembly, which is supported on a circular base by rugged springs at its periphery. When leaves are fed into the assembly through a hopper, oversized particles leaves are immediately discharged through a tangential outlet. Smaller particles pass on down to the next screen to produce finer grades. The machine's horizontal and vertical motion depends on the eccentric weights fitted at the top and bottom. Any increase in the top eccentric weight increases horizontal motion, which increases the machine's output. similarly, more bottom eccentric weight increases vertical motion, which keeps particles in circulation. The tangential motion, meanwhile, is controlled by the weights' angular position. It is at a minimum when the top and bottom weights are lined up; there is then virtually no tendency for the material to spiral. But if the weights are not lined up, the particles spiral from the center out to periphery. In this manner the leaf is sorted, with the finest particles spiralling furthest out.

Hammer Mill

- The Hammer Mill: The Hammer Mill is used to produce dust teas in a continuous manner. It is normally installed apart from other sorting machines so that the dusty atmosphere can be minimized in the sorting room.

Important parts:

- Grinding chamber and wheel: The grinding chamber, which holds the grinding wheel, is made of thick steel. The latter is made of a cluster of carbon steel strikers running at very high speeds of up to 4,500 rpm. The size of the machine is expressed according to its grinding wheel - i.e., 6, 8, 10, 15, 18 or 20 inch.

- Collecting chamber: This is a cylindrical chamber made of steel plate supported by three rods. Dust tea outlets through a vent at the chamber's bottom.
- Motor: The motor of the Hammer Mill runs at 3,000 rpm.
- Working principle: When the coarse product of other sorting machines is fed into the Hammer Mill's grinding chamber through the hopper, it comes into contact with the grinding wheel revolving at 4,500 rpm. The leaf is reduced into a fine power by the striker's sharp cutting edges. The dust tea so produced is directed by a suction fan into a collecting chamber and is directly discharged into the container. The velocity of the dust at the collecting chamber's inlet can be minimized by enlarging the duct through which it travels.

Capacity and power required: The capacity of the Hammer Mill varies according to its size. The average capacities of and power required by various Hammer Mills are:

Size (inch)	Capacity (kgs/hr)	Power (HP)
6	65	5
7	140	7.5
10	160	12.5
15	200	20
18	500	30
20	1,000	40

- Advantage: The greatest advantage of the Hammer Mill is its simplicity and versatility.
- Disadvantage: Its high power requirement is the Hammer Mill's biggest drawback.

Andrew Breaker

In 1948, a machine was developed for breaking tea. It is called the Andrew Breaker, after its developer H.C. Andrew. Today, this machine is an integral of the modern tea sorting room. Earlier machines used for breaking tea could not solve a critical problem -i.e. when leaf is broken into small grades, so is the stalk and this complicates its removal. The Andrew Breaker eliminated this problem in sorting.

Important parts:

- Roller: The most important part of the Andrew Tea Breaker consists of a pair of fluted rollers. The rollers - of 11 inch diameter and 17.5 length - run in mesh but are not in contact. The rollers rotate at 60 rpm and are spring loaded. A lever is provided to adjust both the gap between the engaged teeth of the rollers as well as the pressure applied to the tea particles passing through. Three types of rollers are used in the Andrew Breaker. The first kind of roller is the longitudinal fluted roller. These are sub-divided into two types - the

8 and 16 fluted. types. Eight-fluted longitudinal rollers are general purpose rollers. The increased number of closer flutes in the 16-fluted longitudinal rollers are used in CTC manufacture for even grading and appearance. The third type of roller is also 8-fluted. But here the roller is fluted circumferentially because of which it is known as the corrugated type. This is a special purpose roller used for the treatment of broken teas or stalky coarse residue.

- Working principle: When rough tea is fed to the Andrew Breaker (fitted with longitudinal fluted rollers) through a conveyor system, it is compressed between the rollers rotating at 60 rpm. The amount of tea broken depends on the degree of pressure applied and the speed at which it is applied. Care must be exercised here to optimise the speed at which pressure is applied; that way there is no damage to the natural bloom of the made tea. After extracting all normal, worthwhile grades from the bulk, stalky coarse residue can be fed to the Andrew Breaker fitted with corrugated rollers. In these rollers, the flutes run around the periphery of the rollers. These rollers cause a rubbing action that removes the skin from stalk and tends to disintegrate the tough and flaky particles that previously escaped breakage. The speed of the rollers here is maintained at 180 rpm.
- Capacity: An Andrew Breaker can deal with 3,000 kgs of rough tea per hour with thick and continuous feed.
- Power requirement: The machine requires 3 HP to run.

TEA LEAF GRADE TERMINOLOGY

- Choppy - When the tea contains a lot of varying sized leaves.
- Fannings - Small particles of tea leaves a grade higher than dust. Used almost exclusively in tea bags.
- Flowery - A large leaf, typically plucked in second of third flush, with an abundance of tips
- Golden Flowery - The tea contains very young tips or buds (which are usually golden in colour) that were picked early in the season.
- Tippy - The tea contains an abundance of tips.

WHOLE LEAF GRADES

The grades for whole leaf orthodox black tea (in ascending order) are:

- OP - Orange Pekoe - Main grade in tea production. Can consist of long wiry leaf without tips.
- OP sup - Orange Pekoe Superior - Primarily from Indonesia, tea is much the same as above
- F OP - Flowery Orange Pekoe - High quality tea with a long leaf and few tips, considered second grade in Assam, Dooars and Bangladesh and first grade in China

- F OP1 - Flowery Orange Pekoe First Grade Leaves - As above but with only the highest quality leaves in the F.O.P classification
- GF OP1 - Golden Flowery Orange Pekoe First Grade Leaves- Higher proportion of tip than FOP Top grade in Milima and Marinyn regions; Uncommon in Assam and Darjeeling.
- TGF OP - Tippy Golden Flowery Orange Pekoe - Tea with the highest proportion of tip; Main grade in Darjeeling and Assam.
- TGF OP1 - Tippy Golden Flowery Orange Pekoe - As above, but with only the highest quality leave in the T.G.F.O.P classification
- FTGF OP - Finest Tippy Golden Flowery Orange Pekoe - Highest quality grade. Often hand processed and produced at only the best plantations. Roughly one quarter tips.

A joke among tea aficionados is that "FTGFOP" stands for "Far Too Good For Ordinary People".

BROKEN LEAF GRADES

- BT - Broken Tea - Usually a black, open, fleshy leaf, very bulky. Classification used in Sumatra, Sri Lanka, some parts of Southern India.
- BP - Broken Pekoe- Most common broken pekoe grade. From Indonesia. Ceylon, Southern India.
- BPS - Broken Pekoe Souchong - Term for broken pekoe in Assam and Darjeeling.
- FP - Flowery Pekoe - High quality pekoe. Usually coarser, fleshier broken leaf. from Ceylon and Southern India, also produced in some parts of Kenya.
- BOP - Broken Orange Pekoe - Main broken grade. Prevalent in Ceylon, Southern India, Java and China.
- F BOP - Flowery Broken Orange Pekoe - Coarser broken with some tips from Assam, Ceylon, Indonesia, China and Bangladesh. In South America coarser, black broken.
- F BOP F - Finest Broken Orange Pekoe Flowery - The finest broken orange pekoe. Higher proportion of tips. Mainly from Ceylon's "low districts".
- G BOP - Golden Broken Orange Pekoe - Second grade tea, uneven leaves and few tips.
- GF BOP1 - Golden Flowery Broken Orange Pekoe 1 - As above, but with only the highest quality leaves in the GFBOP classification.
- TGF BOP1 - Tippy Golden Flowery Broken Orange Pekoe 1 - High quality leaves with high proportion of tips. Finest broken First Grade Leaves in Darjeeling and some parts of Assam.

FANNINGS GRADES

- PF - Pekoe Fannings -

- OF - Orange Fannings - From Northern India and some parts of Africa and South America.
- FOF - Flowery Orange Fannings - Common in Assam, Dooars and Bangladesh. Some leaf sizes come close to the smaller broken grades.
- GFOF - Golden Flowery Orange Fannings- Finest grade in Darjeeling for teabag production.
- TGFOF - Tippy Golden Flowery Orange Fannings.
- BOPF - Broken Orange Pekoe Fannings - Main grade in Ceylon, Indonesia, Southern India, Kenya, Mozambique, Bangladesh and China. Black-leaf tea, few added ingredients, uniform particle size, no tips.

DUST GRADES

- D1 - Dust 1 - From Sri Lanka, Indonesia, China, Africa, South America and Southern India.
- PD - Pekoe Dust
- PD1 - Pekoe Dust 1 - Mainly produced in India.

6

Tea Tasting

Tea tasting is the process in which a trained taster determines the quality of a particular tea. Due to climatic conditions, topography, manufacturing process, and different clones of the Camellia sinensis plant (tea), the final product may have vastly differing flavours and appearance. These differences can be tasted by a trained taster in order to ascertain the quality prior to sale or possibly blending tea.

The ISO 3103 standard describes a standardised method for brewing teas in order to make meaningful sensory comparisons between them. It is not a particularly useful standard for many teas, however, requiring a six-minute brewing time and boiling water, neither of which are recommended for green teas (usually brewed at < 90 °C and for under 3 minutes). Tea tasters would instead usually taste teas which have been brewed according to the recommendations of the producer.A tea taster uses a large spoon and noisily slurps the liquid into his/her mouth - this ensures that both the tea and plenty of oxygen is passed over all the taste receptors on the tongue to give an even taste profile of the tea. The liquid is then usually spat back out into a spitoon before moving onto the next sample to taste. The flavour characteristics and indeed leaf colour, size and shape are graded using a specific language created by the tea industry to explain the overall quality. Generally speaking, once the quality has been tasted/graded, each tea company places a value on it based on market trends, availability and demand.

HOW TO TASTE TEA

But like a good wine, chocolate or coffee, tea is the product of its environment, how the tea is grown and how it's processed. These variables make each tea unique and give it body, character, aroma (and health properties). Here at Twinings, our professional tea tasters - who blend and craft your favourite cuppa - have spent years building their tasting vocabulary. So, for us mere 'tea drinkers' it is going to take us a while to develop ours. The following guide, is just that, it's designed to help you study your teas taste, flavour and aroma. We are hoping by using this guide that then you can tell us what you think of our new teas as you discover them, with you becoming our trusted, valued friends that we can then call on, in the future, during the development process of creating new tea blends before they are fully developed.

PPEARANCE, AROMA, FLAVOUR AND MOUTH FEE

hen we're tasting tea we look for four key features: appearance, aroma, flavour and mouth feel. You will learn about all of these on this tea tasting journey... Interestingly, their roots lie in the Chinese tea ceremony. Understanding the full picture of what is happening around you is one of their key teachings.

First Impressions Count

As a tea taster, the first thing you do is inspect the dry tea. Surprisingly, a large proportion of our overall option is based on its look:

- Shape - what are the bits of tea like? Small? Large?
- Colour - is it all the same colour? Is it jet black? If it's all different colours then it might be a blend
- Texture - is it finely ground like dust? Is it crumbly? Sticky?
- Tea Buds - can you see any?

Signs of a good loose tea:

- In black, white and greem tea: gold or bronze flecks - these are the young leaves called tippy
- It should crunch between your fingers - this means it hasn't absorbed any moisture (or flavour) from the air around it
- Buds

Making the Perfect Brew

We've already started to talk about set standards - tea tasters talk about them a lot. They're necessary to make sure that every tasting is consistent. So we'll always use the same.

- Standard Approach by Tea Tasters - this is somebody who tastes tea the same way.
- Weight of tea - between 2 and 3 grams in the UK (a heaped teaspoon).
- Water - Use what you would normally drink - whether that's from the tap or a bottle.
- Rockery - Every tea taster in the world will use the same crockery - but you can just find the purest, whitest crockery you can get your hands on. This will help you to see the colours and depths.
- Brew time - that's three and a half minutes. It's the right amount of time for the colour, flavour and goodness to come out.

Its best not to overly squeeze your tea bag because this could release deep rooted tannins and they taste very bitter

Now what does the Infused Tea Look Like?

In the world of tea, we call the infusion 'liquor'. When we're inspecting the liquor, we look for.....

- Colour - does it leap out at you? We are looking for a bright, jewel-like colour.
- Physical Appearance - we are looking for shiny, slightly oily and bright? It's fine to have a little fragments floating around the bottom of the cup
- Clarity - Once we've brewed the tea, we look at the wet leaves. For Twinngs Whole Leaf Silky pyramids, why not carefully break open the tea bag and take a closer look? What do you see? What can you smell? We're looking for a great colour. And to see that the leaves have become plump and whole again.

By now you're probably dying to try your tea, so let's move onto the taste...

How do we Taste? Here's the Science Bit

When we think of taste, we think of our tongue. Some studies have suggested that up to 90% of flavour is perceived through smell. Our tongues detect five essential tastes and these give us our initial impression:

- When we think of taste, we think of our tongue. But in fact up to 90% of flavour is perceived through smell.
- Our tongues detect five essential tastes and these give us our initial impression sweet, salty, acidic, bitter,umami (a word from Japanese, meaning 'pleasant savoury taste'). The initial perception of something can sometimes throw you off the taste, so that's why it's important to first take in the aroma.
- Our taste buds help form part of an intricate system that allows our brain to decide on a taste. The taste buds along with gustatory receptors and the Olfactory gland allow our brain soon makes a fast decision on if as an individual we recognise and like the flavour of what we are about to ingest
- The Olfactory Gland is situated several centimetres behind the back of our eyes and nose. With its fine hairs on its surface it captures molecules of what we are smelling and in some case putting in our mouth (that's why it's important to slurp our tea to mix with the air).
- A combination of the neural messages from the tongue, olfactory gland and with some help from the gustatory receptors we are quickly able to build a profile of the tea we are drinking.

Now let's put our olfactory gland to the test...

Aroma - Let's Take a Little Sniff

Because we mainly taste through our sense of smell, it's important to deeply consider the aroma. There are two techniques for sniffing, these are:

- Deep inhalations - this is when you hold the brew as close to your nose as possible and take a deep breath.
- Dog action - this is when you take rapid, shallow inhalations through the nose, rather like a panting dog.
- Use your taste wheel to help you consider some of the aromas.
- At this point you will start to get the first persception of the flavour.

How to Taste and Slurp

Here's the fun bit. It's when we actually get to taste the tea. So throwing all table manners out of the window - here's what you do:

- Scoop up some of the liquor with your spoon.
- Take deep breath (first).
- Pucker up like you're about to give someone a kiss, then slurp the liquid up into your mouth from the surface of the spoon. The louder the slurp, the better. You do this to mix oxygen with the liquor as it helps to bring the flavours to life.

Now you need to breath out through your nose (whilst keeping your mouth closed) - this is called retro-olfaction perception. Then swallow the liquor.

- Pay special attention to any sensations created on the tongue for example sweetness or savoury. And remember that bitterness is present in the majority of teas because of varying degrees of tannins.
- Now that you've mastered the art of slurping, let's move on to inspecting the taste.

What's the Flavour Like?

When you're tasting not all aromas and flavours are detectable at the same time. There can be many complex layers of taste for us we will just look for three different notes: head, body and tail. You can use our flavour wheel for this part. Head notes (First impressions) are the ones that give you the initial impression, and they come thick and fast.

Thinking about this first impression look at the inner ring of your taste wheel and describe what you feel you taste, different people may find different tastes, after a few years of tasting in this way you will find that you start to taste in a standard way, as we discussed before. Then you feel the body notes (secondary ring). These give you the overall lasting impression and character. And finally, the tail notes (after taste) are the ones that linger and stay with you after the liquid has passed from your mouth - look at the outer ring. This final stage often indicates the complexity and, in some cases, the quality. Have a go at describing the flavour using your taste wheel. Here's an example of how it might sound: "A spicy head, a nutty body and a hazelnut tail".It's funny to talk about mouth feel, but it's important as it indicates strength. Again, we can use the taste wheel for this but it's often better to use what feels right.

Mouth feel is all about the sensations you feel in your mouth when you taste tea. And of course, different teas trigger different sensations. Some are smooth and round, some are drying and bite into the jaw - but all of them help us to decide if the tea feels right. Often this is the connection between taste and smell - it's what makes us want to taste the tea again.

THE ART OF TEA TASTING

Many of us taste tea everyday, but how often do we take the time to truly discover each of the aromas that hit our nose, and the flavors that meet our palette? The complexity of flavor that comes with high quality tea approaches the complexity of a robust wine. Unraveling the incredible variety of tastes, textures and aromas of your favorite teas is a fun way to enjoy the adventure of drinking teas that come from all over the world.

And yet, tasting tea and finding ways to describe what meets your tongue can be daunting. Experienced tea tasters from tea gardens around the world taste thousands of cups of tea every year. Here at two leaves and a bud tea company, we're tasting different "crops" of tea weekly, carefully noting subtle differences in cup color, aroma, texture and flavor profiles.

Where are you, the average "cuppa' day" tea drinker, supposed to begin? Fear not—the following step-by-step guide to the art of tea tasting will have you tasting and describing teas like a connoisseur in no time. Let's begin!

Properly preparing the tea you're about to taste is crucial. Interestingly, preparing tea for a formal tasting is quite different from your normal home preparation. Here's what you'll want to do: Your average tea sachet or cup of loose tea is going to hold around 2.8 grams of tea, steeped in 350 mL of water. For your tea tasting, measure out double the amount of tea you would normally sip, or just use half the water – 175 mL.

We'll explain why in the next section. eat your water. Here, you'll want to be very careful about what temperature your water reaches. Basic guidelines are: White and green teas: these delicate leaves can't handle too much heat, or else they end up singed with a bitter, burnt flavor. Steep at temperatures between 170 and 185 degrees Fahrenheit.

- Oolong teas: add a little more heat! Brew at temperatures between 180 and 190 degrees Fahrenheit.
- Black and herbal teas: Heartier leaves can handle a full boil. Steep at a temperature between 208 and 212 degrees to release the full flavor.

Start steeping! Pour your water over the leaves and cover the top of your mug or teapot to preserve the heat. Set a timer for 4-5 minutes (a full-fledged, long steep), and remove the tea after the timer has gone off.

All done? You're now ready to begin tasting!

You're probably asking yourself – why did I steep double the amount of tea I would normally? The answer: so that your cuppa' tea "overemphasizes" the flavors hidden in each tea. A full, long steep with double the amount of tea might not be perfectly pleasing to the palate, but it will help to tease out the flavors you might not notice normally, and make them more apparent to your tongue.

With that in mind, there are also a few techniques during the actual tasting itself that will help bring out the flavors of the tea and make them more recognizable: Smell: Studies have shown that 80-90% of what we taste is actually due to what we smell. With that in mind, you'll want to treat your cuppa' tea just like a glass of wine: take a deep, long inhale as you bring the tea to your mouth, so that the aroma mingles with what you're tasting on your tongue.

Slurp! It may make your tea tasting a bit more rude and noisy, but slurping tea is the only way to truly taste it. Slurping cools the tea, and runs it over your tongue, allowing it to fully touch every part of your palette. And don't swallow just yet! You'll want to let the tea sit on your tongue while slightly pulling in air from your lips (cue more slurping sounds), thus aerating the tea and allowing it to fully run over your palette, bringing out the full range of flavor.

Spit: Now your tea tasting just got really rude! Spitting out your tea like you would at a wine tasting is purely optional, but many tea tasters spit out their tea so that they don't become overly "jived" from caffeine. Wondering if you need to cleanse your palette when you're tasting multiple teas? You actually don't. Tea tasters transition from tea to tea without pause to quickly compare flavors. Don't worry—the flavor of the previous tea won't "dilute" what's in your mouth now.

Now comes the hard part: taking what you've tasted and putting it into words. The disclaimer for this step is that it is purely subjective – there is no "right" answer. What you taste is what you taste, and any adjective you'd like to use to describe your tea is just fine.

- Black Teas: The earthy, deep tones of many black teas will often fall under the Wood and Mineral categories of the flavor wheel.
- Green Teas: Green teas taste…green. The vegetal tasting profile of green tea means that it generally falls under the Herbal section of the flavor spectrum.
- Oolong Teas: Oolong's flavor can be smoky and/or sweet, green and/

or black. It often occupies a middle ground in the realm of tea between green and black teas, and so its flavor profile can run the gamut of Spice, Sweet, Wood, and Mineral.

- White Teas: Incredibly subtle in their flavor, white teas are light, bright, and often have a little sweetness. You'll often find pleasant Floral aspects in white tea, along with flavors in the Sweet realm of the spectrum.
- Flavored and Herbal Teas: The above tea profiles are used to describe the flavor of the actual tea itself. But what if you have a Jasmine Petal green tea? Or an herbal blend like our Organic Better Belly Blend with eleven different herbs? The answer is simple – look to the ingredients! If you're sipping two leaves and a bud's Jasmine Petal green tea, combine descriptors of green tea with some jasmine petal descriptors from the "floral" or "sweet" category. If you're drinking our Organic Mountain High Chai Black Tea, which contains spices like cinnamon and clove, combine descriptors of black tea with words from the "spice" category.

TOOLS AND MATERIALS

- A few selections of tea that you want to study and compare (it is wise to compare selections from the same category each round, esp in the beginning, before you are really very confident in the process)
- Taster's Mugs, 150 ml capacity, coupled with tea bowls for holding the infused liquor
- Tasting Spoons
- Bowl for warming the spoons
- Bowl for used spoons
- Tea Holders
- Water Kettle, preferably electric ones with 2000W heating capability or above, 1 litre capacity or above, or Instant Tankless Water Boiler
- Kitchen thermometer
- Electronic Scale, 0 ~ 20 g
- Electronic Timer
- White bread
- (optional) Warm milk, cheese, or soda crackers just in case there is a little stomach un-comfort for over-drinking

As a hobbyist or a connoisseur, you do not need to line up your kitchen table with 50 taster's mugs and slurp from them all day long. 5 or 6 is a good number to start with. In fine teas, where aroma and taste are critically associated with temperature, one really cannot do more than 10 at a time anyway. Party up with your tea friends so you have more teas to play with in each session, if you want.

PREPARATION

- Measure 3 grams of each tea and put inside the leaf-holders in a line ready for use (note the sequence so you know which tea is which)
- Preheat mugs in a steamer, oven, or by filling with boiling water for a couple of minutes
- Line the mugs in a straight line, with enough hand space in between, say 4 inches (10 cm)
- Put the measured tea in the mug, in the same sequence as the line up
- Set the timer to the required infusion time
- Fill the mug, to the rim, consecutively with water at the temperature designated for the specific tea, timing the intervals between the filling of each mug
- Cover the lid immediately after filling of each mug
- Let steep for the required time, 5 or 6 minutes, but have to be the same for each mug
- While steeping, put a tea bowl in front of each mug
- Warm the tasting spoons in a separate bowl of hot water
- When the first mug is infused to the needed time, decant by holding tight the lid and letting the liquor drain through the slits into the bowl. Time the decanting for the next mug to the exact time lapse during step 05, such that each mug is infused for the same amount of time. Go through the line to decant all the mugs
- (optional step)Afterdecanting each mug, invert the mug so the infused leavesfallonto, and piled on the mug lid. Place the lid upside down,withtealeaves showing,backonto the mug. This is done so you can study the infused leaves

TASTING

- Make sure your palate is clean.
- Starting from the tea which you think will be the lightest taste (not necessarily the lightest colour), sniff the tea bowl. (or sniff the inside of the mug lid, if you have not chosen to perform step 12)
- Scoop a spoonful to slurp in (with air, please don't mind the manner) and let it roll in your tongue. Let the liquid rest a second and roll it again before swallowing.
- Feel your palate, tongue, throat and stomach for their reactions with the drink.
- Halt for 5 to 10 seconds before you go for the next mug, after taking a small bite of bread to clean your palate and clear the lingers of the previous tea. Remember to clear the palate every time when you go back to cross compare different selections

- When the infused leaves are a bit cooler, sniff them. Compare the appearance and smells
- Keep notes for each selection for later reference.

ASK YOURSELF

- How does the liquor smell?
- How does the inside of the mug lid smell?
- Does the aroma linger? Which part of it lingers longer? Which shorter?
- Does any part of the aroma remind you of another substance?
- How does the liquor taste?
- Does any part of the taste come first and some other later?
- What do the different parts of the taste remind you of?
- Which part of your tongue gives you the sensation of the taste?
- How does the taste stay?
- Does the taste change during the process of tasting?
- What is the texture of the tea liquor?
- Do you feel any taste at the throat? Or elsewhere?
- How do the infused leaves smell? Same as the liquor? Or different?
- What are the characteristic differences between different selections?
- How does your stomach feel a few seconds after drinking? A few minutes?
- Do you feel any other thing?

There are professionals in the trade who rather spit out the tea in each slurp so they do not have to intake a vast amount of tea, because some really do this whole day long. For people in the fine tea business, however, the differences a tea can make at the throat and after swallowing are quite important in judging a tea. Sometimes it makes a really big difference. Therefore, I suggest people who care to give up mass market quality to experience the complete process of drinking a fine tea. After a couple of sessions and a number of selections, you'll begin to appreciate which are the finer teas and teas that you personally prefer. However, since your taste buds and your body needs time to adjust to fine tea, and each individual can be very different in this, do give yourself some time and more opportunities. Remember, price really should not interfere with your judgment of a selection. As a trader myself, I give all my samples serial numbers and hide all other information such as growers or price, so I do not have any prejudice when tasting a tea.

TEA TASTING TERMINOLOGY

- Bloom: A sign of good manufacture and sorting(where the reduction of leaf has mainly taken place before firing). A "sheen" which has not been removed by over-handling or over-sorting.

- Black: A black appearance is desirable, preferably with "bloom': This term is used with Orthodox or Rotorvane manufacture.
- Blackish: This is a satisfactory appearance for CTC and LTP manufacture teas, and denotes careful sorting.
- Brown: A brown appearance, with CTC and L TP manufacture, normally reflects too hard treatment of the leaf.
- Bold: Particles of leaf which are too large for the particular grade.
- Clean: Leaf which is free from fibre, dust or any extraneous matter.
- Stock & Fibre: Should be minimal in primary or top grades, but generally unavoidable in the lower grades at an assortment.
- Grainy: Describes well made CTC or L TP primary grades, more particularly Dusts.
- Even: True to the grade and consisting of pieces of leaf of quite even size.
- Uneven & Mixed: "Uneven" pieces of leaf usually indicative of poor sorting and not true to the particular grade.
- Flaky: Flat, open and often light in texture.
- Grey: Caused by too much abrasion during sorting.
- Leafy: Orthodox manufacture leaf tending to be on the large or long side.
- Light: A tea light in weight and of poor density. Sometimes flaky.
- Make: Well made (or not) and must be true to the grade.
- Neat: A grade having good "make" and size.
- Nose: Smell of the dry leaf.
- Choppy: Orthodox (or Rotorvane) manufacture leaf which has to be cut by a "breaker" during sorting.
- Chesty: Inferior or unseasoned packing materials cause this taint.
- Powdery: Fine light dust.
- Tip: A sign of fine plucking and apparent in the top grades' of Orthodox manufacture.
- Well Twisted: Applicable to Orthodox manufacture. Often referred to as "well made" or "rolled' and used for describing whole leaf grades.
- Bright: A lively bright appearance. Usually indicates bright liquors
- Dull: Lacks brightness and usually denotes a poor tea. Can be due to faulty manufacture and firing, or a high moisture content.
- Green: Caused by under-fermentation, or characteristic of leaf from immature bushes (liquors often raw or light)
- Mixed or Uneven: Leaf of varying colour
- Aroma: Smell or scent denoting "inherent character"; usually at high elevations.
- Brisk: The most "live" characteristic. Results from good manufacture.
- Bright: Denotes a lively fresh tea with good keeping quality.

- Coloury: Indicates useful depth of colour and strength.
- Character: An attractive taste when describing better high elevation growth, and peculiar to origin.
- Flavour: A most desirable extension of "character" caused by slow growth at high elevations and comparatively rare.
- Strength: Substance in cup.
- Quality: Refers to "cup quality" and denotes a combination of the most desirable liquoring properties.
- Cream: A precipitate obtained after cooling.
- Full: A good combination of strength and colour.
- Pungent: Astringent with a good combination of briskness brightness and strength. (More related to best quality North Indian teas).
- Dry: Indicates slight over-firing.
- High-Fired: Over-fired but not bakey (or burnt).
- Bakey: An over-fired liquor.
- Burnt: Extreme over-firing.
- Dull: Not clear, and lacking any brightness or briskness.
- Fruity: Can be due to over-fermentation and/or bacterial infection before firing. An over-ripe taste.
- Earthy: Normally caused by damp storage. A taste which can at times be "climatically inherent" in leaf from certain origins.
- Gone Off: A flat or old tea. Often denotes a high moisture content.
- Harsh: A taste generally related to under-withered leaf.
- Light: Lacking strength and any depth of colour.
- Green (or RAW) : An immature character. Often due to underfermentation (and sometimes under-wither).
- Plain: A liquor which is "clean" but lacking in the desirable characteristics.
- Coarse: Fibre content.
- Baggy: A taint normally resulting from unlined hessian bags.
- Flat: Unfresh, (usually due to age).
- Smoky: Mainly caused by leaks around the dryer heating tubes.
- Soft: The opposite of briskness and lacking any "live" characteristic. Caused by inefficient fermentation and/or firing.
- Stewed: A soft liquor with an undesirable taste Caused by faulty firing at low temperatures and often insufficient air flow.
- Thin: An insipid light liquor which lacks any desirable characteristics.
- Weedy: A grass or hay taste related to under-withering.
- Taints (in general): A character or taste which is "foreign" to tea.

7

Tea Plant Diseases

BACTERIAL DISEASES

XANTHOMONAS CAMPESTRIS

Xanthomonas campestris is bacterial species that causes a variety of plant diseases. Available from the NCPPB and other international culture collections such as ICMP, ATCC, and LMG in a purified form, it is used in the commercial production of a high-molecular-weight polysaccharide - xanthan gum - that is an efficient viscosifier of water and that has many important uses, especially in the food industry. It causes spots on the infected plant.

Types of Xanthomonas Campestris

- Xanthomonas campestris pv. armoraciae
- Xanthomonas campestris pv. begoniae A
- Xanthomonas campestris pv. begoniae B
- Xanthomonas campestris pv. campestris
- Xanthomonas campestris pv. carota
- Xanthomonas campestris pv. corylina
- Xanthomonas campestris pv. dieffenbachiae
- Xanthomonas campestris pv. graminis
- Xanthomonas campestris pv. hederae
- Xanthomonas campestris pv. hyacinthi
- Xanthomonas campestris pv. juglandis - the walnut blight
- Xanthomonas campestris pv. malvacearum or Xanthomonas citri subsp. malvacearum
- Xanthomonas campestris pv. musacearum
- Xanthomonas campestris pv. nigromaculans
- Xanthomonas campestris pv. pelargonii
- Xanthomonas campestris pv. phaseoli
- Xanthomonas campestris pv. poinsettiicola
- Xanthomonas campestris pv. prunii

- Xanthomonas campestris pv. raphani
- Xanthomonas campestris pv. sesami
- Xanthomonas campestris pv. tardicrescens
- Xanthomonas campestris pv. translucens
- Xanthomonas campestris pv. vesicatoria
- Xanthomonas campestris pv. viticola

The former X. c. pv. citri, which causes citrus canker, was reclassified as X. axonopodis in 1995. In 2006, the species designations for X. c. pv. citri and X. c. pv. malvacearum were revised to X. citri and these pathovars are now referred to as subspecies.

PSEUDOMONAS AVELLANAE

Pseudomonas avellanae is a Gram-negative plant pathogenic bacterium. It is the causal agent of bacterial canker of hazelnut (Corylus avellana). Based on 16S rRNA analysis, P. avellanae has been placed in the P. syringae group. This species was once included as a pathovar of Pseudomonas syringae, but following DNA-DNA hybridization, it was instated as a separate species. Following ribotypical analysis Pseudomonas syringae pv. theae was incorporated into this species.

AGROBACTERIUM TUMEFACIENS

Agrobacterium tumefaciens (updated scientific name: Rhizobium radiobacter) is the causal agent of crown gall disease (the formation of tumours) in over 140 species of eudicots. It is a rod-shaped, Gram-negative soil bacterium. Symptoms are caused by the insertion of a small segment of DNA (known as the T-DNA, for 'transfer DNA'), from a plasmid, into the plant cell, which is incorporated at a semi-random location into the plant genome. A. tumefaciens is an alphaproteobacterium of the family Rhizobiaceae, which includes the nitrogen-fixing legume symbionts. Unlike the nitrogen-fixing symbionts, tumor-producing Agrobacterium species are pathogenic and do not benefit the plant. The wide variety of plants affected by Agrobacterium makes it of great concern to the agriculture industry. Economically, A. tumefaciens is a serious pathogen of walnuts, grape vines, stone fruits, nut trees, sugar beets, horse radish, and rhubarb.

Conjugation

To be virulent, the bacterium must contain a tumour-inducing plasmid (Ti plasmid or pTi), of 200 kb, which contains the T-DNA and all the genes necessary to transfer it to the plant cell. Many strains of A. tumefaciens do not contain a pTi. Since the Ti plasmid is essential to cause disease, prepenetration events in the rhizosphere occur to promote bacterial conjugation - exchange of plasmids amongst bacteria. In the presence of opines, A. tumefaciens produces a diffusible conjugation signal called 30C8HSL or the Agrobacterium autoinducer. This activates the transcription factor TraR, positively regulating the transcription of genes required for conjugation.

Method of Infection

A. tumefaciens infects the plant through its Ti plasmid. The Ti plasmid integrates a segment of its DNA, known as T-DNA, into the chromosomal DNA of its host plant cells. A. tumefaciens has flagella that allow it to swim through the soil towards photoassimilates that accumulate in the rhizosphere around roots. Some strains may chemotactically move towards chemical exudates from plants, such as acetosyringone and sugars. The former is recognised by the VirA protein, a transmembrane protein encoded in the virA gene on the Ti plasmid. Sugars are recognised by the chvE protein, a chromosomal gene-encoded protein located in the periplasmic space.

At least 25 vir genes on the Ti plasmid are necessary for tumor induction. In addition to their perception role, virA and chvE induce other vir genes. The virA protein has autokinase activity: it phosphorylates itself on a histidine residue. Then the virA protein phosphorylates the virG protein on its aspartate residue. The virG protein is a cytoplasmic protein produced from the virG Ti plasmid gene. It is a transcription factor, inducing the transcription of the vir operons. The chvE protein regulates the second mechanism of the vir genes' activation. It increases VirA protein sensibility to phenolic compounds.

Attachment is a two-step process. Following an initial weak and reversible attachment, the bacteria synthesize cellulose fibrils that anchor them to the wounded plant cell to which they were attracted. Four main genes are involved in this process: chvA, chvB, pscA, and att. The products of the first three genes apparently are involved in the actual synthesis of the cellulose fibrils. These fibrils also anchor the bacteria to each other, helping to form a microcolony.

VirC, the most important virulent gene, is a necessary step in the recombination of illegitimate recolonization. It selects the section of the DNA in the host plant that will be replaced and it cuts into this strand of DNA.

After production of cellulose fibrils, a calcium-dependent outer membrane protein called rhicadhesin is produced, which also aids in sticking the bacteria to the cell wall. Homologues of this protein can be found in other rhizobia.

Possible plant compounds that initiate Agrobacterium to infect plant cells:

- Acetosyringone and other phenolic compounds
- Alpha-Hydroxyacetosyringone
- Catechol
- Ferulic acid
- Gallic acid
- p-Hydroxybenzoic acid
- Protocatechuic acid

- Pyrogallic acid
- Resorcylic acid
- Sinapinic acid
- Syringic acid
- Vanillin

Formation of the T-pilus

To transfer the T-DNA into the plant cell, A. tumefaciens uses a type IV secretion mechanism, involving the production of a T-pilus. When acetosyringone and other substances are detected, a signal transduction event activates the expression of 11 genes within the VirB operon which are responsible for the formation of the T-pilus.

The pro-pilin is formed first. This is a polypeptide of 121 amino acids which requires processing by the removal of 47 residues to form a T-pilus subunit. The subunit is circularized by the formation of a peptide bond between the two ends of the polypeptide.

Products of the other VirB genes are used to transfer the subunits across the plasma membrane. Yeast two-hybrid studies provide evidence that VirB6, VirB7, VirB8, VirB9 and VirB10 may all encode components of the transporter. An ATPase for the active transport of the subunits would also be required.

Transfer of T-DNA into the Plant Cell

The T-DNA must be cut out of the circular plasmid. A VirD1/D2 complex nicks the DNA at the left and right border sequences. The VirD2 protein is covalently attached to the 5' end. VirD2 contains a motif that leads to the nucleoprotein complex being targeted to the type IV secretion system (T4SS).

In the cytoplasm of the recipient cell, the T-DNA complex becomes coated with VirE2 proteins, which are exported through the T4SS independently from the T-DNA complex. Nuclear localization signals, or NLSs, located on the VirE2 and VirD2, are recognised by the importin alpha protein, which then associates with importin beta and the nuclear pore complex to transfer the T-DNA into the nucleus. VIP1 also appears to be an important protein in the process, possibly acting as an adapter to bring the VirE2 to the importin. Once inside the nucleus, VIP2 may target the T-DNA to areas of chromatin that are being actively transcribed, so that the T-DNA can integrate into the host genome.

Genes in the T-DNA

Hormones

To cause gall formation, the T-DNA encodes genes for the production of auxin or indole-3-acetic acid via the IAM pathway. This biosynthetic pathway is not used in many plants for the production of auxin, so it means the plant has no molecular means of regulating it and auxin will be produced constitutively. Genes for the production of cytokinins are also expressed. This stimulates cell proliferation and gall formation.

Opines

The T-DNA contains genes for encoding enzymes that cause the plant to create specialized amino acids which the bacteria can metabolize, called opines. Opines are a class of chemicals that serve as a source of nitrogen for A. tumefaciens, but not for most other organisms. The specific type of opine produced by A. tumefaciens C58 infected plants is nopaline.

Two nopaline type Ti plasmids, pTi-SAKURA and pTiC58, were fully sequenced. A. tumefaciens C58, the first fully sequenced pathovar, was first isolated from a cherry tree crown gall. The genome was simultaneously sequenced by Goodner et al. and Wood et al. in 2001. The genome of A. tumefaciens C58 consists of a circular chromosome, two plasmids, and a linear chromosome. The presence of a covalently bonded circular chromosome is common to Bacteria, with few exceptions. However, the presence of both a single circular chromosome and single linear chromosome is unique to a group in this genus. The two plasmids are pTiC58, responsible for the processes involved in virulence, and pAtC58, dubbed the "cryptic" plasmid.

The pAtC58 plasmid has been shown to be involved in the metabolism of opines and to conjugate with other bacteria in the absence of the pTiC58 plasmid. If the pTi plasmid is removed, the tumor growth that is the means of classifying this species of bacteria does not occur.

Beneficial Uses

The DNA transmission capabilities of Agrobacterium have been vastly explored in biotechnology as a means of inserting foreign genes into plants. Marc Van Montagu and Jeff Schell, (University of Ghent and Plant Genetic Systems, Belgium) discovered the gene transfer mechanism between Agrobacterium and plants, which resulted in the development of methods to alter the bacterium into an efficient delivery system for genetic engineering in plants. The plasmid T-DNA that is transferred to the plant is an ideal vehicle for genetic engineering. This is done by cloning a desired gene sequence into the T-DNA that will be inserted into the host DNA.

This process has been performed using firefly luciferase gene to produce glowing plants. This luminescence has been a useful device in the study of plant 'chloroplast' function and as a reporter gene. It is also possible to transform Arabidopsis thaliana by dipping flowers into a broth of Agrobacterium: the seed produced will be transgenic. Under laboratory conditions, the T-DNA has also been transferred to human cells, demonstrating the diversity of insertion application. The mechanism by which Agrobacterium inserts materials into the host cell by a type IV secretion system is very similar to mechanisms used by pathogens to insert materials (usually proteins) into human cells by type III secretion. It also employs a type of signaling conserved in many Gram-negative bacteria called quorum sensing. This makes Agrobacterium an important topic of medical research, as well.

FUNGAL DISEASES

- Cercoseptoria ocellata: Cercoseptoria ocellata is a fungal plant pathogen that affects the tea plant (Camellia sinensis).
- Cercospora theae: Cercospora theae is a fungal plant pathogen.
- Pseudocercospora theae: Pseudocercospora theae is a fungal plant pathogen infecting tea.
- Cylindrocladium lanceolatum: Cylindrocladium lanceolatum is a fungal plant pathogen.
- Rosellinia arcuata: Rosellinia arcuata is a plant pathogen infecting tea.
- Rosellinia bunodes: Rosellinia bunodes is a plant pathogen infecting several hosts including avocados, bananas, cacao and tea.
- Corticium invisum: Corticium invisum is fungus that is a plant pathogen.
- Corticium theae: Corticium theae is fungus that is a plant pathogen.
- Exobasidium vexans: Exobasidium vexans is a plant pathogen affecting tea. Blister blight caused by Exobasidium vexans is a devastating leaf disease in tea (Camellia sinensis) in almost all tea growing regions in Asia. This disease causes serious crop losses under inclement weather conditions besides affecting quality of made tea. Although tea cultivars show varying degrees of resistance/susceptibility to blister blight, a cultivar showing total resistance to blister blight has not yet been identified. The mature two-celled basidiospores are very easily dislodged from the sterigmata and are usually to be found on the surface of the blister. In carefully collected material we have seen the mature two-celled basidiospores attached to sterigmata. Although the basidiospore is normally one-septate, as many as three septa have been seen in germinated spores. The view is expressed that the extra septa are normally formed during germination.
- Lasiodiplodia theobromae: Lasiodiplodia theobromae is a plant pathogen with a very wide host range. It causes rotting and dieback in most species it infects. It is a common post harvest fungus disease of citrus known as stem-end rot. It is a cause of bot canker of grapevine. It also infects Caesalpinia sappan, a species of flowering tree also known as Sappanwood. On rare occasions it has been found to cause fungal keratitis, lesions on nail and subcutaneous tissue. It has been implicated in the widespread mortality of Baobab (Adansonia

digitata) trees in Southern Africa. A preliminary study found the deaths to have a complex set of causes requiring detailed research.

- Glomerella cingulata: Glomerella cingulata is a plant pathogenic fungus that causes disease on many different hosts including quince and apple bitter rot and anthracnose on many fruit and vegetable species such as mango or cultivated plants like St. John's Wort (Hypericum perforatum). It also causes leaf spot on Hevea brasiliensis. Danielone is a natural phenol and a phytoalexin produced by the papaya Carica papaya to fight Colletotrichum gloeosporioides. Glomerella cingulata is the sexual stage (teleomorph) while the asexual stage (anamorph) is called Colletotrichum gloeosporioides.
- Colletotrichum coccodes: Colletotrichum coccodes is a plant pathogen, which causes anthracnose on tomato and black dot disease of potato. Fungi survive on crop debris and disease emergence is favored by warm temperatures and wet weather
- Phellinus noxius: Phellinus noxius is a plant pathogen.
- Calonectria colhounii: Calonectria colhounii is a fungal plant pathogen.
- Ceuthospora lauri: Ceuthospora lauri is a plant pathogen.
- Kretzschmaria deusta: Kretzschmaria deusta, commonly known as brittle cinder, is a fungus and plant pathogen found in temperate regions of the Northern Hemisphere. It is common on a wide range of broadleaved trees including beech (Fagus), oak (Quercus), lime (Tilia), and maple (Acer). It causes a soft rot, breaking down both cellulose and lignin, and decays the trunk and/or roots of living trees. The fungus continues to decay wood after the host tree has died, making K. deusta a facultative parasite. The resulting brittle fracture has a ceramic-like fracture surface. Black zone lines can be seen in cross-sections of wood infected with K. deusta. New fruiting bodies are formed in the spring and are flat and gray with white edges. The fruiting bodies persist all year and their appearance changes to resemble asphalt, consisting of black, domed, lumpy crusts. The mature fruiting bodies are brittle like charcoal and are usually found between the root buttresses.
- Kretzschmaria zonata: Kretzschmaria zonata is a plant pathogen.
- Phomopsis theae: Phomopsis theae is a plant pathogen infecting tea.
- Guignardia camelliae: Guignardia camelliae is a plant pathogen.
- Calonectria kyotensis: Calonectria kyotensis is a fungal plant pathogen.
- Athelia (fungus): Athelia is a genus of corticioid fungi in the family Amylocorticiaceae. Some species are facultative parasites of plants (including crops) and of lichens. The widespread genus contains 28 species.

- Phomopsis theae: Phomopsis theae is a plant pathogen infecting tea.
- Nectria cinnabarina: Nectria cinnabarina is a plant pathogen that causes cankers on many tree species and also a disease known as coral spot. A study published in 2011 showed that this species is actually a complex consisting of at least 4 distinct species.
- Pestalotiopsis theae: Pestalotiopsis theae is a plant pathogen affecting tea.
- Pestalotiopsis longiseta: Pestalotiopsis longiseta is a plant pathogen infecting tea.
- Marasmius crinis-equi: Marasmius crinis-equi is a plant pathogen. It is commonly known as the 'horse hair fungus', and appears on rainforest leaves as a wiry stipe with a delicate fruitbody. The cap of the fruitbody can be up to 4 mm in diameter, and is pale brown.
- Calonectria pyrochroa: Calonectria pyrochroa is a fungal plant pathogen infecting tea.
- Calonectria ilicicola: Calonectria ilicicola is a fungal plant pathogen.
- Calonectria indusiata: Calonectria indusiata is a fungal plant pathogen with several hosts.
- Hendersonia theicola: Hendersonia theicola is a fungal plant pathogen.
- Pestalotiopsis adusta: Pestalotiopsis adusta is a plant pathogen infecting tea and avocados.
- Phaeosphaerella theae: Phaeosphaerella theae is a fungal plant pathogen infecting tea.
- Pleospora theae: Pleospora theae is a plant pathogen infecting tea.
- Elsinoë theae: Elsinoe theae is a plant pathogen.
- Macrophoma theicola: Macrophoma theicola is a plant pathogen affecting Camellia sinensis, the tea plant, which is capable of reducing yields or killing plants altogether. Early symptoms include the browning and drooping of leaves. It is spread most easily in wet, rainy conditions.
- Exobasidium reticulatum: Exobasidium reticulatum is a plant pathogen.
- Laetiporus baudonii: Laetiporus baudonii is a plant pathogen.
- Phyllosticta erratica: Phyllosticta erratica is a plant pathogen infecting tea.
- Phyllosticta theae: Phyllosticta theae is a plant pathogen infecting tea.
- Phanerochaete salmonicolor: Phanerochaete salmonicolor is a plant pathogen infecting several hosts.
- Poria hypobrunnea: Poria hypobrunnea is a plant pathogen infecting tea.
- Helicobasidium compactum: Helicobasidium compactum is a plant pathogen.

- Ganoderma philippii: Ganoderma philippii is a plant pathogen infecting cacao, tea and coffee trees.
- Schizopora flavipora: Schizopora flavipora is a plant pathogen infecting tea.
- Cephaleuros virescens: Cephaleuros virescens is a plant pathogen infecting tea, coffee and coconut.
- Gliocladiopsis tenuis: Gliocladiopsis tenuis is a fungal plant pathogen. It is found in Vietnam.
- Cylindrocladiella camelliae: Cylindrocladiella camelliae is a fungal plant pathogen that causes root rot in tea.
- Cylindrocladiella camelliae: Cylindrocladiella camelliae is a fungal plant pathogen that causes root rot in tea.
- Cylindrocladium clavatum: Cylindrocladium clavatum is a fungal plant pathogen.
- Fomes lamaënsis: Fomes lamaënsis is a plant pathogen of the tea plant.
- Diplodia theae-sinensis: Diplodia theae-sinensis is a plant pathogen.
- Capnodium footii: Capnodium footii is a sooty mold that develops in coconut leaves.
- Capnodium theae: Capnodium theae is a plant pathogen that affects the leaves of the tea plant.
- Irpex destruens: Irpex destruens is a species of fungus in the family Meruliaceae. A plant pathogen, it causes stump rot. It was first described scientifically by English mycologist Thomas Petch in 1909.
- Biscogniauxia capnodes: Biscogniauxia capnodes a species of fungus in the family Xylariaceae.It is a plant pathogen.
- Tunstallia aculeata: Tunstallia aculeata is an ascomycete fungus that is a plant pathogen infecting tea.
- Marasmius tenuissimus: Marasmius tenuissimus is a species of agaric fungus in the large agaric genus Marasmius.
- Septobasidium bogoriense: Septobasidium bogoriense is a plant pathogen, one of a number of fungi in the genus Septobasidium responsible for the disease of tea plants known commonly as "velvet blight".
- Septobasidium pilosum: Septobasidium pilosum is a plant pathogen, one of a number of fungi in the genus Septobasidium responsible for the disease of tea plants known commonly as "velvet blight".
- Septobasidium theae: Septobasidium theae is a plant pathogen, one of a number of fungi in the genus Septobasidium responsible for the disease of tea plants known commonly as "velvet blight".
- Nectria mauritiicola: Nectria mauritiicola is a fungal plant pathogen infecting cacao trees.

- Rigidoporus microporus: Rigidoporus microporus is a plant pathogen, known to cause white root rot disease on various tropical crops, such as cacao, cassava, tea, with economical importance on the para rubber tree (Hevea brasiliensis).
- Elsinoë leucospila: Elsinoë leucospila is a species of fungus in the Elsinoaceae family. A plant pathogen, it was first formally described in 1946.
- Sphaceloma theae: Sphaceloma theae is a plant pathogen infecting tea.
- Phyllosticta theicola: Phyllosticta theicola is a plant pathogen infecting tea.
- Nemania serpens var. serpens: Nemania serpens var. serpens is a plant pathogen.
- Nemania diffusa: Nemania diffusa is a plant pathogen.

ANTHRACNOSE

- Gloeosporium theae-sinensis: Gloeosporium theae-sinensis (syn. Colletotrichum theae-sinensis) is a plant pathogen.
- Gloeosporium theae-sinensis: Gloeosporium theae-sinensis (syn. Colletotrichum theae-sinensis) is a plant pathogen.

ARMILLARIA ROOT ROT

Armillaria Fuscipes

Armillaria fuscipes is a plant pathogen that causes Armillaria root rot on Pinus, coffee plants, tea and various hardwood trees. It is common in South Africa. The mycelium of the fungus is bioluminescent.

Armillaria Heimii

Armillaria heimii is a species of fungus in the Physalacriaceae family of mushrooms.

Armillaria Mellea

Armillaria mellea, commonly known as honey fungus, is a basidiomycete fungus in the genus Armillaria. It is a plant pathogen and part of a cryptic species complex of closely related and morphologically similar species. It causes Armillaria root rot in many plant species and produces mushrooms around the base of trees it has infected. The symptoms of infection appear in the crowns of infected trees as discoloured foliage, reduced growth, dieback of the branches and death.

The mushrooms are edible but some people may be intolerant to them. This species is capable of producing light via bioluminescence in its mycelium.

Armillaria mellea is widely distributed in temperate regions of the Northern Hemisphere. The fruit body or mushroom, commonly known as stump mushroom, stumpie, honey mushroom, pipinky or pinky, grows typically on hardwoods but may be found around and on other living and dead wood or in open areas. The species was originally named Agaricus melleus by Danish-Norwegian botanist Martin Vahl in 1790; it was transferred to the genus Armillaria in 1871 by Paul Kummer. Numerous subtaxa have been described:

Name	Authority	Year
var. viridiflava	Barla	1887
var. minor	Barla	1887
var. bulbosa	Barla	1887
var. camerunensis	Henn.	1895
var. exannulata	Peck	1893
var. flava	Peck	1897
var. glabra	Gillet	1874
var. javanica	Henn.	1900
var. laricina	(Bolton) Barla	1887
var. maxima	Barla	1887
var. obscura	Gillet	1874
var. radicata	Peck	1891
var. sulphurea	(Weinm.) Fr.	1879
var. tabescens	(Scop.) Rea & Ramsb.	1917
var. versicolor	(With.) W.G.Sm.	1908
subsp. nipponica	J.Y.Cha & Igarashi	1995
f. rosea	Calonge & M.Seq.	2003

Description

The basidiocarp of each has a smooth cap 3 to 15 cm (1 to 6 in) in diameter, convex at first but becoming flattened with age often with a central raised umbo, later becoming somewhat dish-shaped. The margins of the cap are often arched at maturity and the surface is sticky when wet. Though typically honey-coloured, this fungus is rather variable in appearance and sometimes has a few dark, hairy scales near the centre somewhat radially arranged. The gills are white at first, sometimes becoming pinkish-yellow or discoloured with age, broad and fairly distant, attached to the stipe at right angles or are slightly decurrent. The spore print is white. The stipe is of variable length, up to about 20 cm (8 in) long and 3.5 cm (1.4 in) in diameter. It is fibrillose and of a firm spongy consistency at first but later becomes hollow. It is cylindrical and tapers to a point at its base where it is fused to the stipes of other mushrooms in the clump. It is whitish at the upper end and brownish-yellow below, often with a very dark-coloured base. There is a broad persistent skin-like ring attached to the upper part of the stipe. This has a velvety margin and yellowish fluff underneath and extends outwards as a white partial veil protecting the gills when young. The flesh of the cap is whitish and has a sweetish odour and flavour with a tinge of bitterness. Under the microscope, the spores are approximately elliptical, 7–9 by 6–7 μm, inamyloid with prominent apiculi (short, pointed projections) at the base. The basidia (spore-producing structures) lack basal clamps.

The main part of the fungus is underground where a mat of mycelial threads may extend for great distances. They are bundled together in rhizomorphs that are black in this species. The fungal body is not bioluminescent but its mycelia are luminous when in active growth.

Similar Species

Armillaria mellea once included a range of species with similar features that have since been reclassified.

Distribution

Armillaria mellea is widespread in northern temperate zones. It has been found in North America, Europe and northern Asia, and It has been introduced to South Africa. The fungus grows parasitically on a large number of broadleaf trees. It fruits in dense clusters at the base of trunks or stumps.

Ecology

Trees become infected by Armillaria mellea when rhizomorphs growing through the soil encounter uninfected roots. Alternatively, when infected roots come into contact with uninfected ones the fungal mycelium may grow across. The rhizomorphs invade the trunk, growing between the bark and the wood and causing wood decay, growth reduction and mortality.

Trees that are already under stress are more likely to be attacked but healthy trees may also be parasitized. The foliage becomes sparse and discoloured, twig growth slows down and branches may die back. When they are attacked, the Douglas-fir, western larch and some other conifers often produce an extra large crop of cones shortly before dying. Coniferous trees also tend to ooze resin from infected areas whereas broad-leaved trees sometimes develop sunken cankers. A growth of fruiting bodies near the base of the trunk confirms the suspicion of Armillaria root rot.

In 1893, the American mycologist Charles Horton Peck reported finding Armillaria fruiting bodies that were "aborted", in a similar way to specimens of Entoloma abortivum. It was not until 1974 that Roy Watling showed that the aborted specimens included cells of both Armillaria mellea and Entoloma abortivum.

He thought that the Armillaria was parasitizing the Entoloma, a plausible hypothesis given its pathogenic behaviour. However, a 2001 study by Czederpiltz, Volk and Burdsall showed that the Entoloma was in fact the microparasite. The whitish-grey malformed fruit bodies known as carpophoroids were the result of E. abortivum hyphae penetrating the Armillaria and disrupting its normal development.

Edibility

Armillaria mellea mushroom are considered good edibles, although some individuals have reported "allergic" reactions that result in stomach upsets. Some authors suggest not collecting mushrooms from the wood of various trees, including hemlock, buckeye, eucalyptus, and locust. The mushrooms have a taste that has been described as slightly sweet and nutty, with a texture ranging from chewy to crunchy, depending on the method of preparation. Parboiling mushrooms before consuming removes the bitter taste present in some specimens, and may reduce the amount of gastrointestinal irritants. Drying the mushrooms preserves and intensifies their flavour, although reconstituted mushrooms tend to be tough to eat.

Chemistry

Several bioactive compounds have been isolated and identified from the fruit bodies. The triterpenes 3β-hydroxyglutin-5-ene, friedelane-2α,3β-diol, and friedelin were reported in 2011. Indole compounds include tryptamine, L-tryptophan and serotonin.

XYLARIA

Xylaria is a genus of ascomycetous fungi commonly found growing on dead wood. The name comes from the Greek xýlon meaning wood. Two of the common species of the genus are Xylaria hypoxylon and Xylaria polymorpha. Xylaria hypoxylon, known by the common names stag's horn and candle-snuff fungus, is the most conspicuous because of its erect, 3–7 cm tall, antler-like ascocarps (fruitbodies) which are black at the base (where the perithecia are embedded) but white and branched towards the top, where the fruiting bodies produce white conidia (asexual spores). Xylaria polymorpha, dead man's fingers, often grows in finger-like clusters from the base of a tree or from wood just below ground level. This is a primary fungus utilized in the spalting of sugar maple and other hardwoods. Xylaria longipes, known by the common names dead moll's fingers and pénis de bois mort, allegedly improves the quality of the wood used in string instruments. It has not been linked to spalting of maple.

BISCOGNIAUXIA NUMMULARIA

Biscogniauxia nummularia is a plant pathogen in the family Xylariaceae, known as the beech tarcrust. The specific epithet is derived from the Latin "nummus" meaning a coin, referring to the often rounded and coin-like encrustations.

Description

The fruit body forms a thick and shiny black crust, on beech (Fagus) bark and is found at all times of the year. It is not edible. Young specimens are covered by a light brown outer layer. The spores are black to dark brown.

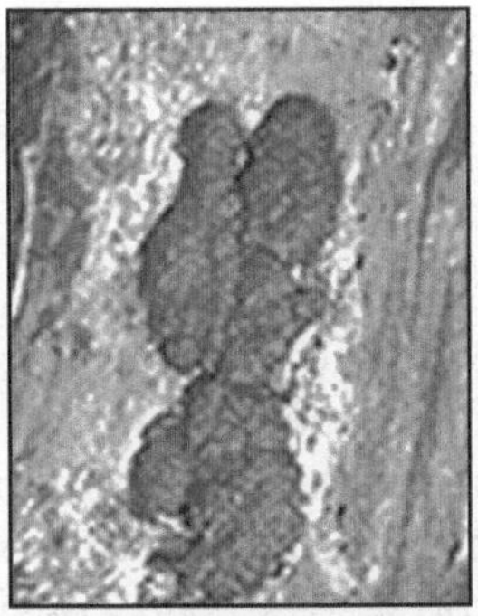

Fig. Detail of the Tarcrust's Structure.

Fig. Beech Tarcrust Growth.

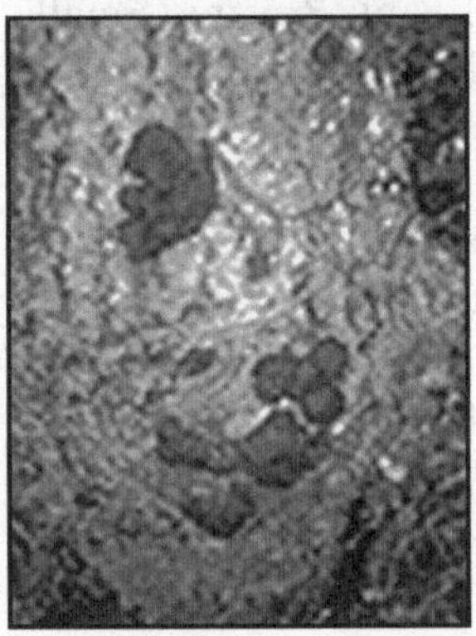

Fig. Encrustations on Beech Bark.

Distribution

Biscogniauxia nummularia is a common fungus specific for Fagus, and has been recorded throughout Europe and Russia, following the extensive distribution of its host tree.

As a Primary Pathogen

The decline of European beech (Fagus sylvatica) in Sicily and Calabria (Italy) has been linked to B. nummularia and experiments have suggested that this ascomycete plays a primary pathogenic role under certain environmental conditions. It typically causes strip-canker and general wood decay.

ATHELIA ROLFSII

Athelia rolfsii is a corticioid fungus in the family Atheliaceae. It is a facultative plant pathogen and is the causal agent of "southern blight" disease in crops. The species was first described in 1911 by Italian mycologist Pier Andrea Saccardo, based on specimens sent to him by Peter Henry Rolfs who considered the unnamed fungus to be the cause of tomato blight in Florida. The specimens sent to Saccardo were sterile, consisting of hyphae and sclerotia. He placed the species in the old form genus Sclerotium, naming it Sclerotium rolfsii. It is, however, not a species of Sclerotium in the strict sense. In 1932, Mario Curzi discovered that the teleomorph (spore-bearing state) was a corticioid fungus and accordingly placed the species in the form genus Corticium. With a move to a more natural classification of fungi, Corticium rolfsii was transferred to Athelia in 1978.

Description

The fungus produces effused basidiocarps (fruit bodies) that are smooth and white. Microscopically, they consist of ribbon-like hyphae with clamp connections. Basidia are club-shaped, bearing four smooth, ellipsoid basidiospores, measuring 4-7 by 3-5 ìm. Small, brownish sclerotia (hyphal propagules) are also formed, arising from the hyphae.

Southern Blight

Athelia rolfsii occurs in soil as a saprotroph, but can also attack living plants. It has an almost indiscriminate host range, but its capacity to form sclerotia (propagules that remain in the soil) means that it particularly attacks seasonal crops. It mostly occurs in warm soils (above 15°C) and can be a serious pest of vegetables in tropical and subtropical regions (including Florida, where it was first recognized), causing "southern blight". It can also be called mustard seed fungus.

LINGZHI MUSHROOM

The lingzhi mushroom or reishi mushroom is a species complex that encompasses several fungal species of the genus Ganoderma, most commonly the closely related species Ganoderma lucidum, Ganoderma tsugae, and Ganoderma sichuanense. G. sichuanense enjoys special veneration in East Asia, where it has been used as a medicinal mushroom in traditional Chinese medicine for more than 2,000 years, making it one of the oldest mushrooms known to have been used medicinally. Lingzhi is listed in the American Herbal Pharmacopoeia and Therapeutic Compendium.

Taxonomy and Naming

Names for the lingzhi fungus have a two thousand year history. The Chinese term lingzhi was first recorded during the Eastern Han dynasty (25–220 CE). Petter Adolf Karsten named the genus Ganoderma in 1881.

Botanical Names

The fungus was given its first binomial name, Boletus lucidus, by English botanist William Curtis in 1781. The lingzhi's botanical names have Greek and Latin roots. The generic name Ganoderma derives from the Greek ganos ãáñïò "brightness; sheen", hence "shining" and derma äÝñìá "skin". The specific epithet lucidum is Latin for "shining." Tsugae is derived from the Japanese word for "hemlock". There are multiple species of lingzhi, scientifically known to be within the Ganoderma lucidum species complex and mycologists are still researching the differences among species within this complex.

Vietnamese Names

The Vietnamese language linh chi is a Chinese loanword used in tieng Viet. It is often used with the Vietnamese word for mushroom nam (nam Linh Chi) which is the equivalent of Ganoderma lucidum or reishi mushroom.

English Names

English lingzhi or ling chih (sometimes spelled "ling chi" from French EFEO Chinese transcription) is a Chinese loanword. The Oxford English Dictionary gives Chinese "líng divine + zhî fungus" as the origin of ling chih or lingzhi, and defines, "The fungus Ganoderma lucidum, believed in China to confer longevity and used as a symbol of this on Chinese ceramic ware." The OED notes the earliest recorded usage of the Wade–Giles romanization ling chih in 1904, and of the Pinyin lingzhi in 1980. In addition to the transliterated loanword, English names include "glossy ganoderma" and "shiny polyporus".

Description

Lingzhi is a polypore mushroom that is soft (when fresh), corky, and flat, with a conspicuous red-varnished, kidney-shaped cap and, depending on specimen age, white to dull brown pores underneath. It lacks gills on its underside and releases its spores through fine pores, leading to its morphological classification as a polypore.

Varieties

With the advent of genome sequencing, the Ganoderma genus has been undergoing taxonomic reclassification. Prior to genetic analyses of fungi, they were classified according to their morphological characteristics. The ITS region of the Ganoderma genome is considered to be a standard barcode marker. It was once thought that Ganoderma lucidum generally occurs in two growth forms- a large specimen with a small or nonexistent stalk found in North America and is sessile, and the other a smaller specimen with a long, narrow stalk found mainly in the tropics. However recent molecular evidence has shown the first form to be a distinct species called G.sessile, a name given to North American specimens by William Alfonso Murrill in 1902. Environmental conditions play a substantial role in the lingzhi's manifest morphological characteristics. For example, elevated carbon dioxide levels result in stem elongation in lingzhi. Other formations include antlers without a cap; these may be affected by carbon dioxide levels as well. The species were also formerly differentiated by their colors, of which the red reishi was considered to be the most researched. The three main factors that most greatly influence fruitbody development morphology are light, temperature, and humidity. (Water and air quality are impactful but to a lesser degree). G. lucidum was previously thought to occupy the Asian and European continents. However, with ITS sequencing, it is now know that G. lucidum sensu stricto is confined to Europe, and G. lucidum comprises two distinct species. Mycologists now consider G. lucidum to be a worldwide distributed species-complex in which the development of each strain is shaped by unique physiological state and metabolic demands.

Biochemistry

Ganoderma lucidum produces a group of triterpenes, called ganoderic acids, which have a molecular structure similar to steroid hormones. It also contains other compounds often found in fungal materials, including polysaccharides (such as beta-glucan), coumarin, mannitol, and alkaloids. Sterols isolated from the mushroom include, ganoderol, ganoderenic acid, ganoderiol, ganodermanontriol, lucidadiol, and ganodermadiol.

Habitat

Ganoderma lucidum, and its close relative Ganoderma tsugae, grow in the northern Eastern Hemlock forests. These two species of bracket fungus have a worldwide distribution in both tropical and temperate geographical regions, growing as a parasite or saprotroph on a wide variety of trees. Similar species of Ganoderma have been found growing in the Amazon. In nature, lingzhi grows at the base and stumps of deciduous trees, especially maple. Only two or three out of 10,000 such aged trees will have lingzhi growth, and therefore its wild form is extremely rare. Today, lingzhi is effectively cultivated both indoors under sterile conditions and outdoors on either logs or woodchip beds.

Preparation

Due to its bitter taste, lingzhi is traditionally prepared as a hot water extract product. Thinly sliced or pulverized lingzhi (either fresh or dried) is added to a pot of boiling water, the water is then brought to a simmer, and the pot is covered; the lingzhi is then simmered for two hours. The resulting liquid is fairly bitter in taste and dark, with the more active red lingzhi more bitter than the black. The process is sometimes repeated for additional concentration. Alternatively, it can be used as an ingredient in a formula decoction or used to make an extract (in liquid, capsule, or powder form). The more active red forms of lingzhi are far too bitter to be consumed in a soup.

Medical Uses

- Anti-Allergic/Anti-Inflammatory Activity: Studies showed that Reishi extract significantly inhibited all four types of allergic reactions, including positive effects against asthma and contact dermatitis and effectively used in treating stiff necks, stiff shoulders, conjunctivitis (inflammation of the fine membrane lining the eye and eyelids), bronchitis, rheumatism and improving "competence" of the immune system without any significant side- effects.
- Anticonvulsant Effects: A water extract from Reishi mycelium significantly increased the threshold for psychomotor seizures in mice. Mycelial extracts also confer anti-inflammatory activity as evidenced by inhibitory activity of lipopolysaccharide (LPS)-induced nitric oxide (NO) production in murine macrophage-like cell line RAW264.7 cells.
- Cancer: The use of G. lucidum has also been explored as a complementary adjunct treatment in patients undergoing chemotherapy treatment. A recent meta-analysis of five randomized control trials showed that patients responded more positively when given G. lucidum alongside their chemotherapy regimen, and the studies also showed that patients had improved immune functions that was measured by their elevated levels of immune response cells. Several compounds in G. lucidum have been studied for apoptotic activity in colon cancer cells, antiproliferative effects in ovarian cancer cells, and induction of apoptosis in human gastric carcinoma cells.
- Cardiovascular Risk Factors: Previous clinical evidence suggested that G. lucidum may have antioxidant, cardioprotective, and glycemic regulatory effects. However, a 2015 Cochrane review did not find evidence to support the use of or treatment of cardiovascular risk factors in people with type 2 diabetes mellitus.
- Diabetes: Several compounds in G. lucidum (including polysaccharides, proteoglycans, proteins and triterpenoids) may have hypoglycemic effects. In vitro evidence suggests that protein tyrosine phosphatase 1B is a promising therapeutic target in diabetes, and a G. lucidum proteoglycan can inhibit this enzyme. Secondly, G. lucidum demonstrates inhibition of aldose reductase and á-glucosidase, which

can suppress postprandial hyperglycemia.A proteoglycan enhanced insulin secretion and decreasing hepatic glucose output (along with increased adipose and skeletal muscle glucose disposal) and normalized serum lipids in a murine model of diabetes. A polysaccharide also demonstrated hypoglycemic effects in type 2 diabetic mice.

- Gastrointestinal Health: Recent murine studies suggest that G. lucidum may positively impact gut microflora to attenuate metabolic risk factors contributing to obesity.
- Hepatoprotection: G. lucidum significantly decreased serum ALT and AST levels in mice livers injured with á-amanitin. A proteoglycan also demonstrated hepatoprotective effects in carbon tetrachloride-induced liver injury in vitro and in vivo.
- Immunostimulation: G. lucidum contains beta glucans and other polysaccharides to stimulate innate immune function and signaling and activate dendritic cells.
- Neuroprotection: G. lucidum protected dopaminergic neurons through inhibition of microglia.

GANODERMA APPLANATUM

Ganoderma applanatum (the Artist's Bracket or Artist's Conk) is a bracket fungus with a cosmopolitan distribution. This fungus grows as a mycelium within the wood of living and dead trees. It forms fruiting bodies that are up to 30–40 centimetres (12–16 in) across, hard, woody-textured, and inedible; they are white at first but soon turn dark red-brown. Brown spores are released from the pores on the underside of the fruiting body. The fruiting bodies are perennial, and may persist for multiple years, increasing in size and forming new layers of pores as they grow.

It is a wood-decay fungus, causing a rot of heartwood in a variety of trees. It can also grow as a pathogen of live sapwood, particularly on older trees. It is a common cause of decay and death of beech and poplar, and less often of several other tree genera, including alder, apple, elm, buckeye and horse chestnut, maple, oak, walnut, willow, western hemlock, Douglas Fir, and Spruce. A peculiarity of this fungus lies in its use as a drawing medium for artists. When the fresh white pore surface is rubbed or scratched with a sharp implement, dark brown tissue under the pores is revealed, resulting in visible lines and shading that become permanent once the fungus is dried. The Anne Frank tree in Amsterdam suffered rot from Ganoderma applanatum before it blew over in 2010. The midge, Agathomyia wankowiczii (Platypezidae) lays its eggs on the fruiting body of the fungus, forming galls. The forked fungus beetle, Bolitotherus cornutus lives out its entire life cycle in the fruiting bodies of Ganoderma applanatum and a few other bracket fungi.

CLADOSPORIUM

Cladosporium is a genus of fungi including some of the most common indoor and outdoor molds. Species produce olive-green to brown or black colonies, and have dark-pigmented conidia that are formed in simple or branching chains. The many species of Cladosporium are commonly found on living and dead plant material.

Some species are plant pathogens, others parasitize other fungi. Cladosporium spores are wind-dispersed and they are often extremely abundant in outdoor air. Indoors Cladosporium species may grow on surfaces when moisture is present. Cladosporium fulvum, cause of tomato leaf mould, has been an important genetic model, in that the genetics of host resistance are understood.

Health effects

Cladosporium species are rarely pathogenic to humans, but have been reported to cause infections of the skin and toenails, as well as sinusitis and pulmonary infections. If left untreated, these infections could turn into respiratory infections like pneumonia.The airborne spores of Cladosporium species are significant allergens, and in large amounts they can severely affect asthmatics and people with respiratory diseases. Cladosporium species produce no major mycotoxins of concern, but do produce volatile organic compounds (VOCs) associated with odours.

COCHLIOBOLUS CARBONUM

Cochliobolus carbonum (anamorph: Helminthosporium carbonum) is one of more than 40 species of filamentous ascomycetes belonging to the genus Cochliobolus (anamorph: Bipolaris/Curvularia). This pathogen has a worldwide distribution, with reports from Australia, Brazil, Cambodia, Canada, China, Congo, Denmark, Egypt, India, Kenya, New Zealand, Nigeria, Solomon Islands, and the United States. Cochliobolus carbonum is one of the most aggressive members of this genus infecting sorghum (Sorghum spp. [Poaceae]), corn (Zea mays [Poaceae]) and apple (Malus domestica [Rosaceae]). As one of the most devastating pathogens of sweet corn, C. carbonum causes Northern leaf spot and ear rot disease while the asexual stage causes Helminthosporium corn leaf spot.

Cochliobolus carbonum is pathogenic to all organs of the corn plant including root, stalk, ear, kernel, and sheath. However, symptoms of infection show distinct manifestations in different plant parts: whole plant - seedling blight affects the whole plant, leaf discoloration and mycelial growth, black fungal spores and lesions appear on inflorescences and glumes, and grain covered with very dark brown to black mycelium which gives a characteristic charcoal appearance due to the production of conidia.

Morphology and Keys to Identification

Cochliobolus carbonum is divided into at least five different races based on pathogenicity, particularly lesion types on corn leaves. Cochliobolus carbonum race 0 (CCR0) is essentially non-pathogenic. Cochliobolus carbonum race 1 (CCR1) produces host specific toxins and oval to circular spots on leaves. While infection by C. carbonum race 2 (CCR2) is rare, it is characterized by the production of oblong, chocolate-colored spots. C. carbonum race 3 (CCR3) produces linear and narrow lesions while C. carbonum race 4 (CCR4) is weakly pathogenic, forming oval to concentric circular spots.

The genus Cochliobolus is distinguished by the presence of dark to black ascomata with a unilocular, globose pseudothecium and a short, cylindrical neck. Ascomata also bear hyphae and conidiophores and asci are bitunicate, 2-8 spored, and cylindrical to obclavate. Ascomata are black, globose to ellipsoidal, and setae are present over the upper half of the wall mixed with conidiophores. Pseudoparaphyses are filiform, hyaline, septate, and branched. The shapes of asci are cylindrical to clavate, short-stalked, straight to slightly curved having 1-8 spores, and vestigial bitunicate.

Conidia are straight to moderately curved, occasionally cylindrical but usually broad in the middle and tapering towards the rounded ends, distoseptate, and 36-100 X 12-18 μm. The surface of the conidia is often granulose and the hilum is inconspicuous. Conidia are produced from the apex of an unbranched conidiophore. Generally, the conidiophore arises singly or in small groups which are straight or flexuous, mid to dark brown, smooth, septate, cylindrical, and up to 250 ìm long, 5-8 ìm thick. Variation in mycotoxin production by Cochliobolus species is used to distinguish some taxa and these mycotoxins are host-specific and non-host specific. HC toxin produced by C. carbonum race 1 and T toxin produced by C. heterostrophus are host-specific toxins while ophiobolins produced by C. miyabeanus and carbotoxin produced by C. carbonum are non-host-specific toxins.

Reproductive Biology

The species most closely related to C. carbonum is C. victoriae which have inter-specific fertility, however only 1% attempted crosses between species were fertile. The sister relationship of these two species were evident based on phylogenetic analysis. Cochliobolus carbonum is the sexual stage and can be obtained by pairing opposite mating single conidial isolates in Sach's agar media with sterilized maize leaf segments or barley grains incubated at 24°C.

Crosses between different species of Cochliobolus are possible, producing offspring differing in conidial morphology. Horizontal and vertical gene transfer has occurred within and between fungal species of Cochliobolus and might be the reason for occurrence of highly virulent, toxin producing races of C. heterostrophus, C. carbonum and C. victoriae. In nature, there is no evidence of gene flow among races of C. carbonum despite their sympatry. The teleomorph of C. carbonum has not been observed in field conditions.

Habitat Preference, Life Cycle, Dispersal and Symptoms Development

Fungi are capable of showing different interactions with their host and different lifestyles depending upon the interaction. Different species of Cochliobolus and its anamorphs are associated with different host species as ephiphytes, endophytes, saprophytes and pathogens. Infected seed is the major source of inoculum of C. carbonum internationally, so it is a quarantined pathogen in Europe and other countries. To my knowledge, few studies have been conducted to better understand the life and infection cycle of C. carbonum. Cochliobolus carbonum survives as mycelium and resistant chlamydospores on maize debris in the field and on infected seed. Conidia serve as a primary source of inoculum dispersed by wind and rain-splash. Damp weather and moderate temperature greatly favor sporulation and produce additional inoculum for secondary spread. During both pathogenic and saprophytic phases of the lifecycle, this fungus enters and ramifies through intact leaves and obtains nutrients from the host cytoplasm and walls by degrading cell wall components through the production of a variety of extracellular enzymes. The symptoms first appear as small, circular to oval, reddish brown to tan lesions and over time become more tan to grayish. In general, moderate temperature, high relative humidity, and heavy dew during the growing season favors the development of this disease. Senescent corn leaves are an important plant part for the growth and development of C. carbonum, because it provides biochemicals required for the formation of perithecia, asci and ascospores.

Important Developments in Classification

Previously, many species of Cochliobolus were placed in Ophiobolus. Drechsler found that graminicolous forms of Helminthosporium produced teleomorphs with characteristic features of greater ascus and ascospore width as well as helicoiod ascospores with bipolar germination, which deviates from previously described characters of Ophiobolus. To accommodate these species, he introduced the new genus Cochliobolus transferred several species to Cochliobolus. Another genus Pseudocochliobolus was separated from Cochliobolus based on the presence of stromatic tissue below the ascomata and the degree of ascospore coiling. The sexual states of Cochliobolus are only found associated with Bipolaris and Curvularia. Bipolaris and Curvularia share some morphological similarities and cannot be easily distinguished by any distinct taxonomic criteria. Although few morphological differences were reported, conidia of Bipolaris are distoseptate while the conidia of Curvularia are bigger and euseptate. Some scientists believe that Bipolaris and Curvularia are synonymous, however there is an existing debate regarding this issue. The different species of fungi belonging to Cochliobolus with asexual states in Bipolaris and Curvularia have undergone frequent name changes as a result of refinement to the taxonomy that has resulted in some confusion. The perfect stage of this fungus was first named as C. carbonum by R. R. Nelson in 1959. The imperfect stage was named as Bipolaris zeicola in 1959. The commonly used synonyms of Cochliobolus carbonum were Helminthosporium zeicola (1930), Helminthosporium carbornum (1944), Drechslera zeiocola (1966) and Drechslera carbonum (1984). Several phylogenetic studies of Bipolaris and Curvularia showed that Bipolaris is not monophyletic and some Bipolaris species are nested within Curvularia.

Phylogenetic Affinities

The taxonomy of Cochliobolus, Bipolaris and Curvularia is confusing due to the frequent name changes and no clear morphological demarcation between Bipolaris and Curvularia. A phylogenetic analysis of Cochliobolus, Bipolaris and Curvularia species was performed using rDNA markers (ITS1, 5.8S, ITS2) and a 600 bp fragment of gpd (glyceraldehyde-3 dehydrogenase) gene. The results showed that most isolates of Cochliobolus and Bipolaris including C. carbonum, that cause serious crop losses, form a separate group from other species. These data revealed that these species radiated form a common ancestor recently. Another phylogenetic study was conducted using a combined analysis of rDNA ITS (internal transcribed spacer), GPDH (glyceraldehyde 3-phosphate dehydrogenase), LSU (large subunit) and EF1-á (translation elongation factor 1-á) and showed that this generic group is divided into two major lineages. According to their study Bipolaris and Cochliobolus species grouped with their type species and Curvularia species, with its generic type, form another group. Similar to previous findings, the analyses showed that Bipolaris and Curvularia cannot be combined into a single monophyletic genus and trees showed both of these groups resolved into single complex. In this paper, the authors claim that they have resolved nomenclatural conflict within this complex based on their phylogenetic data and suggest giving priority to the more commonly used generic names Bipolaris and Curvularia to represent these distinct lineages.

BOTRYTIS CINEREA

Botrytis cinerea ("botrytis" from Ancient Greek botrys (âüôñõò) meaning "grapes" plus the Neolatin suffix -itis for disease) is a necrotrophic fungus that affects many plant species, although its most notable hosts may be wine grapes. In viticulture, it is commonly known as botrytis bunch rot; in horticulture, it is usually called grey mould or gray mold.

The fungus gives rise to two different kinds of infections on grapes. The first, grey rot, is the result of consistently wet or humid conditions, and typically results in the loss of the affected bunches. The second, noble rot, occurs when drier conditions follow wetter, and can result in distinctive sweet dessert wines, such as Sauternes or the Aszú of Tokaji/Grasă de Cotnari. The species name Botrytis cinerea is derived from the Latin for "grapes like ashes"; although poetic, the "grapes" refers to the bunching of the fungal spores on their conidiophores, and "ashes" just refers to the greyish colour of the spores en masse. The fungus is usually referred to by its anamorph (asexual form) name, because the sexual phase is rarely observed. The teleomorph (sexual form) is an ascomycete, Botryotinia fuckeliana, also known as Botryotinia cinerea (see taxonomy box).

Biology

Botrytis cinerea is characterized by abundant hyaline conida (asexual spores) borne on grey, branching tree-like conidiophores. The fungus also produces highly resistant sclerotia as survival structures in older cultures. It overwinters as sclerotia or intact mycelia, both of which germinate in spring to produce conidiophores. The conidia are dispersed by wind and rain-water and cause new infections. A considerable genetic variability has been observed in different Botrytis cinerea strains (polyploidy). Gliocladium roseum is a fungal parasite of Botrytis cinerea.

Viticulture

In the Botrytis infection known as "noble rot" (pourriture noble in French, or Edelfäule in German), the fungus removes water from the grapes, leaving behind a higher percent of solids, such as sugars, fruit acids and minerals. This results in a more intense, concentrated final product. The wine is often said to have an aroma of honeysuckle and a bitter finish on the palate.

A distinct fermentation process initially caused by nature, the combination of geology, climate and specific weather led to the particular balance of beneficial fungus while leaving enough of the grape intact for harvesting. The Chateau d'Yquem is the only Premier Cru Supérieur, largely due to the vineyard's susceptibility to noble rot.

Botrytis complicates winemaking by making fermentation more complex. Botrytis produces an anti-fungal that kills yeast and often results in fermentation stopping before the wine has accumulated sufficient levels of alcohol. Makers of fine German dessert wines have been known to take fermenting tubs of wine into their homes to nurture the yeast through the night to assure that the alcohol level reaches legal minimums for the product to be called wine.

Botrytis bunch rot is another condition of grapes caused by Botrytis cinerea that causes great losses for the wine industry. It is always present on the fruitset, however, it requires a wound to start a bunch rot infection. Wounds can come from insects, wind, accidental damage, etc. To control botrytis bunch rot there are a number of fungicides available on the market. Generally, these should be applied at bloom, bunch closure and veraison (the most important being the bloom application). Some winemakers are known to use the German method of fermentation and prefer having a 5% bunch rot rate in their grapes and will usually hold the grapes on the vine a week longer than normal.

Horticulture

Botrytis cinerea affects many other plants. It is economically important on soft fruits such as strawberries and bulb crops. Unlike wine grapes, the affected strawberries are not edible and are discarded. To minimize infection in strawberry fields, good ventilation around the berries is important to prevent moisture being trapped among leaves and berries. A number of bacteria have been proven to act as natural antagonists to B. cinerea in controlled studies. In greenhouse horticulture, Botrytis cinerea is well known as a cause of considerable damage in tomatoes. The infection also affects rhubarb, snowdrops and white meadowfoam, and cannabis.

Human Disease

Botrytis cinerea mold on grapes may cause "winegrower's lung", a rare form of hypersensitivity pneumonitis (a respiratory allergic reaction in predisposed individuals).

Mycoviruses of Botrytis Cinerea

Botrytis cinerea not only infects plants, it also hosts several mycoviruses itself. A range of phenotypic alterations due to the mycoviral infection have been observed from symptomless to mild impact, or more severe phenotypic changes including reduction in growth/suppression of mycelia, sporulation and sclerotia production, formation of abnormal colony sectors (Wu et al., 2010) and virulence.

RHIZOCTONIA SOLANI

Rhizoctonia solani (teleomorph: Thanatephorus spp.) is a plant pathogenic fungus with a wide host range and worldwide distribution. This plant pathogen was discovered more than 100 years ago. Rhizoctonia solani frequently exists as thread-like growth on plants or in culture, and is considered a soil-borne pathogen. Rhizoctonia solani is best known to cause various plant diseases such as collar rot, root rot, damping off and wire stem. Rhizoctonia solani attacks its host(s) when they are in their juvenile stages of development such as seeds and seedlings, which are typically found in the soil. It makes sense then this saprophytic pathogen would live and survive in the soil, and attack the part of its hosts that reside there. The pathogen is known to cause serious plant losses by attacking primarily the roots and lower stems of plants and although it has a wide range of hosts, their main targets are herbaceous plants. Rhizoctonia

Solani would be considered a basidiomycete fungus if the teleomorph stage were more abundant. The pathogen is not currently known to produce any asexual (conidia) spores, though it is considered to have an asexual life cycle; Occasionally, sexual spores (basidiospores) are produced on infected plants. The disease cycle of Rhizoctonia solani is important in regards to management and control of the pathogen.

In 1858, Julius Kuhn observed a fungus on diseased potato tubers and named it Rhizoctonia solani. This is a widespread, destructive, and versatile plant pathogen. The fungus occurs in all parts of the world, and is capable of attacking many different host plants, causing seed decay, damping-off, stem cankers, fruit decay, and foliage diseases.

Hosts and Symptoms

Rhizoctonia solani causes a wide range of commercially significant plant diseases. It is one of the fungi responsible for Brown patch (a turfgrass disease), damping off in seedlings, as well as black scurf of potatoes, bare patch of cereals, root rot of sugar beet, belly rot of cucumber, sheath blight of rice, and many other pathogenic conditions. The fungus therefore has a wide host range and strains of Rhizoctonia solani may differ in the hosts they are able to infect, the virulence of infection, selectivity for a given host ranges from non-pathogenic to highly virulent, the temperature at which infection occurs, the ability to develop in lower soil levels, the ability to form sclerotia, the growth rate, the survival in a certain area. These factors may or may not show up given the environment and host that Rhizoctonia attacks.

Rhizoctonia solani primarily attacks seeds of plants below the soil surface, but can also infect pods, roots, leaves and stems. The most common symptom of Rhizoctonia is "damping off", or the failure of infected seeds to germinate. Rhizoctonia solani may invade the seed before it has germinated to cause this pre-emergent damping off, or it can kill very young seedlings soon after they emerge from the soil. Seeds that do germinate before being killed by the fungus have reddish-brown lesions and cankers on stems and roots.

There are various environmental conditions that put the plant at higher risk of infection due to Rhizoctonia, the pathogen prefers warmer wet climates for infection and growth. Post-emergent damping off is a further delay in attack of Rhizoctonia solani. The seedling is most susceptible to disease in its juvenile stage.

Cereals in regions of England, South Australia, Canada, and India experience losses caused by Rhizoctonia solani every year. Roots are killed back, causing plants to be stunted and spindly. Other non cereal plants in those regions can experience brown stumps as another symptom of the pathogen. In England, this is called purple patch. Rhizoctonia solani can also cause hypocotyl and stem cankers on mature plants of tomatoes, potatoes and cabbage. Strands of mycelium and sometimes sclerotia appear on their surfaces. Roots will turn brown and die after a period of time. The best known symptom of Rhizoctonia solani is black scurf on potato tubers which are the sclerotia of the fungus.

Disease Cycle

Rhizoctonia solani can survive in the soil for many years in the form of sclerotia. Sclerotia of Rhizoctonia have thick outer layers to allow for survival, and they function as the overwintering structure for the pathogen. In some rare cases(such as the teleomorph) the pathogen may also take on the form of mycelium that reside in the soil as well. The fungus is attracted to the plant by chemical stimuli released by a growing plant and/or decomposing plant residue. The process of penetration of a host can be accomplished in a number of ways. Entry can occur through direct penetration of the plant cuticle/epidermis or by means of natural openings in the plant. Hyphae will come in contact with the plant and attach to the plant by which through growth they begin to produce an appressorium which penetrates the plant cell and allows for the pathogen to obtain nutrients from the plant cell. The pathogen can also release enzymes that break down plant cell walls, and continues to colonize and grow inside dead tissue. This breakdown of the cell walls and colonization of the pathogen within the host is what forms the sclerotia. New inoculum is produced on or within the host tissue, and a new cycle is repeated when new plants become available. The disease cycle begins as such:

- The sclerotia/mycelium overwinter in plant debris, soil or host plants.
- The young hyphae and fruiting basidia (rare) emerge and produce mycelia and rarely basidiospores.
- The very rare production of the germinating basidiospores penetrate the stoma whereas the mycelia land on the plant surface and secrete the necessary enzymes onto the plant surface in order to initiate invasion of the host plant.
- After the mycelia successfully invade the host, necrosis and sclerotia form in and around the infected tissue which then leads to the various symptoms associated with the disease such as soil rot, stem rot, damping off etc. and the process begins all over again.

Environment

The pathogen is known to prefer warm wet weather, and outbreaks typically occur in the early summer months Most symptoms of the pathogen do not occur until late summer and thus most farmers do not become aware of the diseased crop until harvest. A combination of environmental factors have been linked to the prevalence of the pathogen such as: presence of host plant, frequent rainfall/ irrigation and increased temperatures in spring and summer. In addition, a reduction of drainage of the soil due to various techniques such as soil compaction are also known to create favorable environments for the pathogen. The pathogen is dispersed as sclerotia, and these sclerotia can travel by means of wind, water or soil movement between host plants.

Identification

Rhizoctonia solani does not produce spores and is hence identified only from mycelial characteristics or DNA analysis. Its hyphal cells are multinucleate. It produces white to deep brown mycelium when grown on artificial medium. The hyphae are 4–15 µm wide and tend to branch at right angles. A septum near each hyphal branch and a slight constriction at the branch are diagnostic. R. solani is subdivided into anastomosis groups (AG) based on hyphal fusion between compatible strains. The teleomorph of R. solani is Thanatephorus cucumeris. It forms club-shaped basidia with four apical sterigmata on which oval, hyaline basidiospores are borne.

Management

It is not possible to completely control Rhizoctonia, but the severity of the pathogen can be limited. Successful control of Rhizoctonia depends on characteristics of the pathogen, host crops, and environment. Controlling the environment, crop rotation, using resistant varieties, and minimizing soil compaction are effective and noninvasive ways to manage disease. Planting seedlings in warmer soil and getting plants to emerge quickly helps minimize damage. Crop rotation also helps minimize the amount of inoculum that causes Rhizoctonia Solani. There are a few resistant varieties with moderate resistance to Rhizoctonia that can be used, but they produce lower yields and quantity than standard varieties. Minimizing soil compaction is also another way to reduce risk of the pathogen because this helps water infiltration, drainage, and aeration for the plants.

One specific chemical option is a chemical spray PCNB which is known to be the best solution to reducing damping off of seeds on host plants. To minimize disease, we can use plant certified seed that is free of sclerotia. Seed growers should look into only purchasing sclerotia free seeds when planting their crops since sclerotia can overwinter in the soil and may not show symptoms right away. Although fungicides are not the most effective way to manage this pathogen, there have been a few that have been approved by the USDA for control of the pathogen. One should consult their chemical representative on which group of fungicides would be most effective with their crops in regard to Rhizoctonia solani. Avoidance is one of the key ways to ensure that the pathogen will not be among their crops. As long as seed growers stay clear of wet, poorly drained areas while also avoiding susceptible crops, Rhizoctonia solani is not usually a problem. Diseases caused by this pathogen are more severe in soils that are moderately wet and a temperature range of 15-18 degrees C.

Economic Importance

Rhizoctonia can be found across all areas of the United States (environmental conditions permitting) where its host crops are located. The severity of infection can vary and for highly infected patches, severity of the infection cab be very devastating to the farmer. Some of these consequences are : major yield losses(ranging from 25%-100%),increased soil tare(because the soil sticks to the fungus' mycelium), and poor industrial quality of the crops based on increased levels of sodium, potassium and nitrogen. Due to the vast number of hosts that the pathogen attacks, these consequences are numerous and detrimental to a variety of crops.

CERATOBASIDIUM

Ceratobasidium is a genus of fungi in the order Cantharellales. Basidiocarps (fruit bodies) are effused and the genus is sometimes grouped among the corticioid fungi, though species also retain features of the heterobasidiomycetes. Rhizoctonia-like anamorphs of Ceratobasidium species are placed in the genus Ceratorhiza. Species are saprotrophic, but several are also facultative plant pathogens, causing a number of commercially important crop diseases. Some are also endomycorrhizal associates of orchids.

Ceratobasidium was introduced in 1935 by American mycologist D.P. Rogers to accommodate species of the old form genus Corticium that showed affinities with the heterobasidiomycetes. These affinities were the possession of large sterigmata ("Cerato-basidium" means "horned basidium") and the production of basidiospores that produce secondary spores. Four species were originally placed in the genus, with subsequent authors adding a further 35 species.

Current Status

Research on the septal pore ultrastructure of the little-known and atypical type species, Ceratobasidium calosporum, indicates that it is a member of the Auriculariales and is unrelated to other species of Ceratobasidium. This taxonomic problem has not yet been resolved. Molecular research, based on cladistic analysis of DNA sequences, places Ceratobasidium (excluding the type species) within the Cantharellales.

The Genus Ceratorhiza

Many Ceratobasidium species produce anamorphic hyphal states, sometimes with sclerotia, that were originally placed in the form genus Rhizoctonia. With a move to a more natural classification of the fungi, the new genus Ceratorhiza was introduced for anamorphs of Ceratobasidium by R.T. Moore in 1987. Some ten species have been described in Ceratorhiza, several of which are not linked to any known teleomorph.

Description

Fruit bodies are effused, thin and often inconspicuous, smooth, waxy to dry and web-like, whitish to pale grey. Microscopically they have comparatively wide hyphae without clamp connections and basidia that are spherical to cuboid or broadly club-shaped. Basidia bear 2 to 4 sterigmata, which are comparatively large.

Basidiospores are globose to cylindrical (elongated and worm-like in the type species), smooth, and colourless. They frequently produce secondary spores and germinate by hyphal tubes. Ceratorhiza anamorphs produce hyphae (sometimes swollen) and occasionally sclerotia (small propagules composed of thick-walled hyphae).

Habitat and Distribution

Species are mainly saprotrophic, occurring in the soil and producing fruit bodies on dead stems and plant detritus. Some occur on attached leaves and stems. Several species have been isolated from orchid mycorrhiza. Distribution appears to be cosmopolitan.

NEMATODES, PARASITIC

BURROWING NEMATODE

Radopholus Similis

Radopholus similis is a species of nematode known commonly as the burrowing nematode. It is a parasite of plants, and it is a pest of many agricultural crops. It is an especially important pest of bananas and citrus, and it can be found on coconut, avocado, coffee, sugarcane, other grasses, and ornamentals. It is a migratory endoparasite of roots, causing lesions that form cankers. Infected plants experience malnutrition.

History and Distribution

The nematode was first described from necrotic tissue in a species of Musa, the banana genus, in 1891. It is one of the most important root pathogens of banana crops, causing yield losses of up to 30 to 60% in many countries. It is known in temperate regions worldwide.

Morphology

Adults and juveniles are vermiform in shape. Adults are sexually dimorphic. The male has a poorly developed stylet, a knob-like head, and a sharp, curved spicule enclosed in a sac. The male is 500 to 600 μm in length, while the female is about 550 to 880 μm long. The female has a well-developed stylet. Both male and female have long, tapered tails with rounded or indented ends.

Biology

The nematode completes its life cycle in about 21 days at 25°C. Females and juveniles feed inside roots, especially near the tips. Males with their weak stylets do not feed. Females lay two to six eggs per day. The nematode causes a disease condition called toppling or blackhead disease in plants. In bananas they weaken the anchor roots and the plants can fall. The roots also fail to supply the plant with water and nutrients, causing reduced growth and development.

Management

The nematode load in the soil can be reduced with fumigation and crop rotation. Cover crops that are not susceptible to the nematode, such as Crotalaria or Tagetes, can be sown. Disease-free sprouts raised from clean tissue cultures can be used.

DAGGER NEMATODE

- Xiphinema insigne: Xiphinema insigne is a plant pathogenic nematode infecting tea.

LANCE NEMATODE

- Hoplolaimus columbus: Hoplolaimus columbus is a plant pathogenic nematode.

MATURE TEA NEMATODE

Meloidogyne Brevicauda

Meloidogyne brevicauda is a plant parasitic nematode. It is also called Tea root-knot nematode, Mature tea nematode or Indian root-knot nematode. It is a member of the root-knot nematodes, which was identified by C. A. Loos in 1953 in Sri Lanka.

Description

Female M. brevicauda has a narrow procorpus, a large metacorpus, a large glandular region that has one dorsal and two subventral esophageal gland lobes. Two large nucleated esophago-intestinal cells are at the dorsal base of the metacorpus junction with the intestine. The labial disc and lips are prominent and protrude from the regular body contour. The perineal pattern appears with high squared dorsal and ventral arches without prominent lateral lines.

Male M. brevicauda has a slit-like stoma surrounded by six small, pore-like openings of the inner labial sensilla. The head annulations are absent. The lips are set off from the body annulations because the diameter of the lips is larger than that of the first body annule. The first body annule is also narrower than the remaining body annules. The shape of the head of males of M. brevicauda is very different from that of most root-knot males. The labial disc of most males of Meloidogyne is fused with the medial lips, but in M. brevicauda,the labial disc is distantly separated from them by a deep groove. Other than that, most root-knot males have a large head annule that may be further divided by irregular or incomplete annulations. Males of M.brevicauda do not have a head region, the lips are immediately adjacent to the first body annule. The tail is marked by acrescent-shaped fold in the cuticle surrounding the posterior portion of the cloacal opening.

Life Cycle

This root-knot nematode is sedentary endoparasitic nematode. Second-stage juveniles (J2) penetrate host roots where they establish a specialized feeding site (giant cells) in the stele. As J2 develop, they cause root swellings and become swollen females. Females rupture root cortex and sometime protrude with the egg masses from the root surface. J2 emerge from the egg masses and migrate in the soil.

Distribution

Azerbaijan; Fujian Province, China; Tamil Naduand West Bengal, India and Sri Lanka

Hosts

Camellia sinensis (tea) and Crocus sativus (saffron)

Symptoms

Above-ground symptoms on tea are pale, dull leaves or abnormal leaf fall. Infestations are more apparent in bushes recovering from pruning. Severely attacked bushes may fail to recover. The infected roots have large, characteristic galls, many of which have pinhole pits.

Management

- Use nematode-free planting material.
- Apply nematicides in the planting hole in areas with a history of nematode infestationn.
- Avoid susceptible clones.
- Use soil amendments such as refuse tea, compost, castor oil cake and neem oil cake.
- Uproot severely damaged mature tea and replacing it with a nematode trap crop before replanting.

Impact

Although M. brevicauda causes severe damage to tea, there has been no quantification of the damage caused. The risk posed by M. brevicauda is limited due to its limited distribution and host range (tea and saffron).

PIN NEMATODE

- Paratylenchus curvitatus: Paratylenchus curvitatus is a plant pathogenic nematode infecting tea.

RENIFORM NEMATODE

Rotylenchulus Reniformis

Rotylenchulus reniformis, the reniform nematode, is a species of parasitic nematode of plants with a worldwide distribution in the tropical and subtropical regions.

Taxonomy

This nematode has a wide host range, infecting many species of plants around the world. It was first observed on the roots of cowpea in Hawaii, and was described as new species and new genus in 1940. Its specific epithet, reniformis, was inspired by the kidney shape of the adult female. There are now ten species classified in the genus, but R. reniformis is the only species of major economic importance to agriculture.

Distribution and Host Range

R. reniformis has been reported from thousands of localities in the Americas, Africa, Europe, Asia, and Australia. It has a wide host range that includes fruit trees, lentil, cotton, pigeon pea, tea, tobacco, soybean, pineapple, banana, okra, coconut, cabbage, sweet potato, alfalfa, corn, asparagus, palm, cucumber, tomato, squash, cassava, radish, eggplant, guava, melon, and ginger. Certain plants are considered to be non-hosts of the nematode, such as little barley, common barnyard grass, pangola grass, peppers, and some cultivars of black mustard, oat, spinach, and sugarcane.

Morphology

The reniform nematode has esophageal glands overlapping the intestine and a short stylet. The dorsal esophageal gland orifice is located posterior to the stylet knobs. The immature female is slender and may be spiral- or C-shaped in death. It is about 0.3 to 0.5 millimeters long. The mature female has a swollen, kidney-shaped body with a short tail, a short, thin stylet with rounded stylet knobs, a three-part esophagus, a long and narrow isthmus, and a well-developed metacarpus. The vulva is just behind the middle of the body. The male is vermiform: with a wormlike appearance. It has a weak stylet, curved spicules, and a pointed tail. The esophagus is reduced.

Life Cycle

R. reniformis is sedentary semi-endoparasite on the roots of plants. The female penetrates the root and remains in one position at a permanent feeding site with its posterior end projecting from the root. The immature female is the infective agent, attacking the root and growing to maturity at its feeding site. Males and juveniles live in the soil; males are not parasites and do not feed. Under drought conditions the nematode can persist up to two years outside a host by entering an anhydrobiotic state. The life cycle is 17 to 29 days long. The juvenile molts once while still inside the egg. The eggs hatch in 8 to 10 days. The juvenile molts three times to reach the immature stage. The immature female parasitizes the root for one to two weeks. During this time the male deposits sperm, which the female stores until her gonads mature. The nematode can also reproduce via parthenogenesis, without fertilization. Upon maturity the female exits the root and lays up to 200 eggs in a gelatinous matrix.

Host-parasite Relationship

When the immature female penetrates the root, a feeding tube forms from stylet secretions. Nematode infestation can cause symptoms in the plant that resemble those of moisture and nutrient deficiencies. It causes hypertrophy in the pericycle cells of seedling roots and in the periderm cells of the roots of older plants. Root growth slows and secondary root development is limited. Root necrosis has been observed in pineapple and banana. Shoot growth suppression and reduction of fruit quality has been observed in crops such as pineapple. Infested plants can become stunted and chlorotic. Wilt disease can follow when opportunistic fungi such as Fusarium and Verticillium infect the plants, a process observed in cotton. Secondary fungal infection can also cause root decay. Damage from R. reniformis is directly related to the number of nematodes present when the crop is planted. Variation among nematode populations, hosts, environmental conditions, and soil types may alter the threshold or economic injury level across the geographic distribution of the nematode.

Management

Some plants are resistant to this nematode. In susceptible taxa, chemical control is one of the most common management practices. Nematicides before and after planting can be effective. Intercropping and crop rotation, especially with resistant or non-host plants, is used as cultural control to improve soil, increase antagonistic microorganisms, and reduce nematode populations. French marigold (Tagetes patula) and sunn hemp (Crotalaria juncea) are recommended for these uses. The fungus Purpureocillium lilacinum has been used as an agent of biological pest control against the nematode in chickpea.

ROOT-KNOT NEMATODE

Meloidogyne Arenaria

Meloidogyne arenaria is a plant pathogenic nematode. This nematode is also known as the peanut root knot nematode. The word "Meloidogyne" is derived from two Greek words that mean "apple-shaped" and "female". The peanut root knot nematode, M. arenaria is one of the "major" Meloidogyne species because of its worldwide economic importance. M. arenaria is a predominant nematode species in the United States attacking peanut in Alabama, Florida, Georgia, and Texas. The most damaging nematode species for peanut in the USA is M. arenaria race 1 and losses can exceed 50% in severely infested fields. Among the several Meloidogyne species that have been characterized, M. arenaria is the most variable both morphologically and cytologically. In 1949, two races of this nematode had been identified, race 1 which reproduces on peanut and race 2 which cannot do so. However, in a recent study, three races were described (races 1, 2, and 3). López-Pérez et al (2011) had also studied populations of M. arenaria race 2, which reproduces on tomato plants carrying the Mi gene and race 3, which reproduces on both resistant pepper and tomato.

M. arenaria has the same history as the other root knot nematodes. Meloidogyne spp were originally grouped together with the cyst nematodes (Heterodera spp) but they were later separated after they were confirmed to be significantly different from each other. The changes that were made when discovering Meloidogyne spp of nematodes are described below;

- 1855 - Reverend Miles Joseph Berkeley (clergyman) noted galls on cucumbers in greenhouse in England. This was the first official report of nematodes that caused galls on plant roots.
- 1871 - Schmidt described Heterodera schactii the sugarbeet cyst nematode.
- 1872 - Greeff described Anguillula radicicola – a nematode that caused galls on cereals and grasses but this was actually Ditylenchus and not a root knot nematode.
- 1879 - Cornu - described a root-knot nematode as Anguillula marioni .
- 1884 - Muller decided root-knot nematode was the same as Greeff's root-galling nematode and that both should be Heterodera- Heterodera radicicola.
- 1887 - Goeldi - Brazil - described a root knot on coffee - Meloidogyne exigua.
- 1932 - Goodey - decided that use of Heterodera radicicola was incorrect according to International Rules of Zoological Nomenclature as original H. radicicola was not a root-knot nematode. He renamed it Heterodera marioni.
- 1949 – Chitwood removed them from Heterodera because they differed from cyst nematodes. Since the oldest name for the genus was Goeldi's Meloidogyne exigua, that name had precedence.

Chitwood described five species based on perineal patterns, M. arenaria being one of the five.

Distribution of M. Arenaria

Crop damage by this nematode is mostly faced in warmer regions and it is rarely found in areas where the average monthly temperatures approach freezing. The peanut root knot nematode, race 1 is found in all areas where peanuts are grown.

Economic Importance

The peanut root-knot nematode is the most serious nematode pest of the crop and questionably the most serious soil-borne disease problem. For example, the nematode is present in almost all fields where the crop is grown in Florida and economic damage is estimated to occur in greater than 50% of these fields. Individual peanut fields heavily infested with the root-knot nematode have sustained yield losses greater than 75%. In addition to causing considerable damage alone, it also increases the severity and incidence of other soil-borne diseases such as southern stem rot and Cylindrocladium blackrot. M. arenaria is C-rated pest in California.

Morphology and Anatomy of M. Arenaria

Different development stages of this nematode have unique features that are used to describe them. Among the characters commonly used to define nematode species and populations, morphometric characters have a marked significance.

Females are swollen, pear-shaped pearly white and sedentary. They deposit their eggs in a gelatinous mass, which usually protrudes from galled root tissue. Females of this nematode are completely endoparasitic. Mature females of M. arenaria are identified by the cuticular markings in the perineal area, stylet length, punctations on the cuticle. When describing the morphology of M. arenaria race1, Eisenback et al., (1981) reported that in females, the dorsal arch is flattened and rounded. The striae in the arch are slightly indented at the lateral lines and generally form a shoulder on the arch. The stylet is very robust and the cone and shaft are broad. The labial disc and lips are dumb bell shaped and it has large lateral lips that are separated from the medial lips and the head region.

In a study in Argentina, perineal patterns of 100 females from a population of M. arenaria, race 2 growing on tomato plants were analyzed and moderate morphological variations observed. The overall shape was rounded in most cases, but it was oval-shaped in a small proportion (4%). Forty-six percent of females showed low dorsal arch while 38% depicted a high dorsal arc, while the rest were intermediate. Lines in the post-anal region were smooth or wavy, continuous or broken, occasionally forming shoulders (18% of females). Wings were generally not observed. Phasmids were visible in 25% of the patterns and the mean distance between them was 24.33 ± 1.61 mm. Only seven out of one hundred females presented perineal patterns with lateral lines weakly demarcated by forked striae. Adult females were pyriform and lacking a terminal protuberance in the posterior area of the body. The female body lengths ranged between 465 mm and 1,129 mm. The excretory pore of the females was located closer to the base of the stylet than to the median bulb, at approximately 2-stylet lengths, in agreement with a typical M. arenaria specimen. The stylet was dorsally curved.

Second stage juveniles of this nematode are short (400-600 micrometers) with very short stylets (10-15micrometers). The stylet is lightly sclerotized with indistinct Knobs. The cephalic framework is also weakly sclerotized. The esophageal gland overlaps the intestine ventrally and the tail tapers to a pointed tip with a clear terminus In a study, second stage juveniles (J2s) averaged 458.51 ± 20.48 mm in length and 15.35 ± 0.76 mm in width. The rectum was undilated. Tail length (from anus to posterior end) was 56.21 ± 2.78 mm and the tail tip was rounded.

Males of M. arenaria are long (1.0 to 2.0 mm). The stylet is short and has distinct knobs. The esophageal gland overlaps the intestine ventrally. The tail is short and rounded and has no bursa. Spicules open a short distance from the tail tip. The males also have a head cap that is low and slopes posteriorly with 2 or 3 incomplete annulations around the head region. The stylet is pointed and the lumen opening has a small protuberance on the ventral side. From their study, Garcý´a and Sa´nchez-Puerta, (2012) also reported that the mean length of the male body was 1.6 ± 0.31 mm. The head of the males was flat to concave and showed a moderately elevated labial disc. The males also had four lateral lines, which were easily observed.

Symptoms of Root Knot Disease of Peanut

These nematodes cause symptoms both on leaves and roots. Foliar symptoms of root-knot disease may be expressed at any time during the growing season. These symptoms of nematode damage on peanut plants range from stunting, yellowing, wilting and even plant death. Ferris (1999-2012), also reported that nematodes cause yellowing; mid-day wilting; symptoms of water and nutrient stress; and sometimes death, especially if the nematode is interacting with other organisms. They can appear in a peanut crop beginning about 100 days after planting or after periods of dry hot weather. Areas of root-knot nematode damaged peanut are usually round to oblong in shape, and rows of infected plants may never overlap as those of healthy plants. It is common for plants to wilt and eventually die in areas where nematode populations are high.

Root symptoms are similar to those caused by all nematodes including a reduction in number of feeder roots and root stunting, but root knot nematodes cause unique symptoms on roots and pods. Nematodes cause galls (knots) to form on both roots and pods of the peanut, and these galls are highly diagnostic for root-knot nematode presence compared to other nematodes affecting peanut. The presence of galls on roots, pegs and pods appears as single or multiple wart-like growths that may or may not be discolored. As root knot nematode infection progresses, and generally later in the season, secondary root and pod rots cause further damage and eventual death of the plants.

Many soil-borne fungi, especially Sclerotium rolfsii (which causes southern stem rot) and Sclerotinia minor (which causes Sclerotinia blight), will infect the weakened peanut plants and cause more plant death. Root knot nematode infection of the peg weakens the peg so that it breaks during harvest, contributing to additional yield losses because the detached pods are then left in the soil. Infection of the pods leads to a decline in yield quality.

The damage threshold, or the initial nematode population density at which peanut yields begin to decline, has been reported to be in the range of 1 to 10 eggs and second stage juveniles per 500 cm3 soil for M. arenaria on susceptible peanut cultivars.

Reproduction and Life Cycle

Reproduction occurs mitotically by parthenogenesis and the species has chromosome numbers that range from 30 to 50 The life cycle of all Meloidoigyne spp is similar but they have different temperature requirements. At temperatures ranging from 32.8 to 39.50C, M. arenaria reached the egg laying stage in about 21 days on peanut and completed its life cycle in 32 days. The population levels increased faster from a starting level of 10 juveniles per plant that from 1000 or 10,000 per plant.

Adult females deposit eggs into a protective gelatinous matrix, near or just outside the root surface. A single female lays about 500 to 1,500 eggs during her life, which lasts about two to three months. Eggs hatch only under favorable conditions, such as adequate moisture and warm temperatures. Root knot nematode development begins inside the egg. After the completion of embryogenesis, the first-stage juvenile remains inside the egg until it molts into the second stage juvenile. The first stage juvenile (J1) lacks a stylet. Second stage juveniles (J2) hatch from the egg and move freely in the soil in quest for a suitable host plant. Once they find suitable host and enter the root with the help of the stylet, they start to feed within 24 hours, inducing giant cells to form. J2s then molt thrice (to J3, J4 and then the adult) before they mature in to adults. Adult males remain vermiform and get out of the root while females swell and obtain the pyriform/round shape. Females lay eggs in the gelatinous matrix which hatch again and the cycle is repeated if environmental conditions are favorable. The length of the life cycle depends on environmental factors mostly temperature and host status. It may take about 1 month to complete the life cycle. The nematodes also have a very high reproductive rate.

Feeding Behavior of M. Arenaria

Just like the other root knot nematodes, to be able to feed within the host, this nematode must establish an interaction with the plant. J2s are attracted to the root tip in the zone of elongation and the areas of lateral root emergence. Initially it was reported that they are attracted by CO_2, and apparently by small molecules that are dialysable - perhaps amino acids. Recent studies suggest that the attraction may not be to CO_2 but to lower pH resulting from carbonic acid formed from the CO_2 in solution.

The second juvenile (J2) penetrates zone of elongation by mechanical (stylet thrusts) and probably chemical (cellulase and pectinase) means. It moves between, rather than through, cortical cells towards root apex, turns at the meristem, and migrates back to the vascular cylinder in the zone of cell differentiation. Once the J2 is in the root, they migrate intracellularly to the provascular cells and begin to parasitize cells within 24 hours. The nematode initiates formation of the giant cell (also called a nurse cell) from potential vascular tissue. Sub ventral gland become prominent after penetrating the root, and it could be playing a role of secreting enzymes that play a role in the formation of the giant cell which is a large cell with several nuclei. Giant cell are very large in size (100 fold increase), with reduced vacuoles and they are multinucleate containing about 40 to 100 nuclei. The female feeds from the giant cell as it expands further posteriorly accumulating eggs. After formation of the giant cells, gall development occurs though these two events are separate. Galls result from hypertrophy and hyperplasia of the cells around the giant cells. Growth regulators (IAA) are thought to have a role in cell enlargement since they increases cell wall plasticity.

Host Range

M. arenaria parasitizes peanut, vegetables, grasses, fruit ornamentals and tobacco. The root-knot nematode, M. arenaria can infect almost every plant family. While characterizing the host range of the pea nut root knot nematode, López-Pérez et al., 2011, reported that he most frequent hosts of the peanut root knot nematode were vegetables, fruit trees, tobacco, grapevine, and weeds.

Management of M. Arenaria

- Plant resistant cultivars: Although genes conferring resistance to peanut root knot nematode have not been found in cultivated peanut, other Arachis spp. have been identified to be highly resistant or immune to the peanut root knot nematode. Simpson and Starr (1991) and Garcia et al. (1996) reported successful crosses to transfer a high level of nematode resistance into A. hypogaea. The resistance was obtained from a wild species, Arachis cardenasii. Resistance was obtained by introgression of resistance genes into A. hypogaea using a series of backcrosses. Examples of resistant peanut varieties are coan and nematam Unfortunately, both cultivars are highly susceptible to tomato spotted wilt virus and prevalent fungal diseases. In 2008, however, the USDA released a cultivar, Tifguard, which has resistance to both tomato spotted wilt virus and root-knot nematode.

- Rotation with crops that are poor or non-hosts of the nematode. Two year rotations with cotton, bahai grass, or velvet bean are effective also.
- Use of nematicides as needed. Management of root-knot nematodes on peanut has traditionally relied primarily on treatment of infested fields with nematicides such as granular aldicarb (Temik) or the fumigant 1,3-dichloropropene (Telone). Nematicides are usually recommended for high value crops such as peanuts, tobacco and peach
- Plant tolerant varieties. Crop yields will not be affected even in the presence of M. arenaria.

Promising Ongoing Research and other Issues

- Plant breeders and nematologists are actively pursuing the incorporation of the root-knot nematode resistance into varieties adapted to the disease situation in the southeastern U.S. Tifguard is the first of many future varieties that will be released from peanut breeding programs including the UF/IFAS peanut breeding program. These resistant varieties will not solve the peanut root knot nematode problems in peanut production; however, they hold great promise to reduce losses from this disease by more than 50%.
- Since neither coan nor nematem are resistant to the tomato spotted wilt virus (Tospovirus, family Bunyaviridae), efforts to introgress resistance into other high yielding peanut varieties is ongoing. This will reduce crop losses in the areas with high incidences of the virus. The tomato spotted wilt virus, which causes substantial yield losses in Georgia, Alabama, and Florida.

Points to Ponder

- The resistance in Tifguard and other varieties that will soon be released is the same single dominant gene. Continuous planting of peanut with this gene in the same fields is likely to eventually lead to resistance-breaking nematode biotypes. This can be delayed or prevented by utilizing proper crop rotation and even planting non-resistant peanut in the rotation as well.

- Secondly, this resistance does not reduce damage from lesion nematodes which are a major problem in peanut production in the central peninsular production area of Florida

Northern Root-knot Nematode

Northern root-knot nematode (Meloidogyne hapla) is a vegetable pathogen which produces tiny galls (a sore caused by friction and abrasion) on around 550 crop and weed species. They invade root tissue after birth. Females are able to lay up to 1,000 eggs at a time in a large egg mass. By surviving harsh winters, they can survive in cold climates. (Hence, the name, Northern)

Meloidogyne Incognita

Meloidogyne incognita is a nematode (roundworm) in the family Heteroderidae. It is commonly called the "southern root-knot nematode" or the "cotton root-knot nematode". This parasitic roundworm has worldwide distribution and numerous hosts. It is an important plant parasite classified in parasitology as a root-knot nematode, as it prefers to attack the root of its host plant. When M. incognita attacks the roots of plants, it sets up a feeding location, where it deforms the normal root cells and establishes giant cells. The roots become gnarled or nodulated, forming galls, hence the term "root-knot" nematode. M. incognita has been found to be able to move along shallower temperature gradients (0.001C/cm) than any other known organism, an example of thermotaxis. The response is complicated and thought to allow the nematodes to move toward an appropriate level in soil, while they search for chemical cues that can guide them to specific roots.

Meloidogyne Javanica

Meloidogyne javanica is a plant-pathogenic nematode. It is one of the tropical root-knot nematodes and a major agricultural pest in many countries. It has many hosts. Meloidogyne javanica reproduces by obligatory mitotic parthenogenesis (apomixis).

Meloidogyne Thamesi

Meloidogyne thamesi (Thames' root-knot nematode) is a plant pathogenic nematode (roundworm) infecting tea.

ROOT LESION NEMATODE

Pratylenchus Brachyurus

Pratylenchus brachyurus is a plant parasitic nematode.Pratylenchus brachyurus, like many other plant-parasitic nematodes, are microscopic worms that can be damaging to many agricultural crops. Across the world, many billions of dollars each year are lost due to the damaging abilities of nematodes on cash crops. P. brachyurus is a migratory endoparasite that enters the roots and moves throughout the root tissue while completing its life cycle. Under the old classification system based on morphology, these nematodes were under the Class Adenophorea. This changed in 2004 when a new classification system was presented by De Ley and Blaxter using SSU rDNA for comparison between different species of nematodes. Using this new system, De Ley and Blaxter placed P. brachyurus in the Class Chromadorea, Order Rhabditida. In 1929 Godfrey first described P. brachyurus as the main cause for pineapple-root disease and called them Tylenchus brachyurus. He also discussed several other nematodes which were all later put into the Pratylenchus genus. In 1934, Filipjev proposed the genus Pratylenchus. This proposed new genus name came with some challenges because the International Code of Zoological Nomenclature states that the new genus name must come with a statement as to why you want to change the name. It must also give reasons differentiating the new genus plus give the name of the type species. In 1936 the new genus name was universally accepted.

Description

This nematode is a migratory endoparsite that ranges from about 0.4 to 0.5 mm long and has a lip region that is angular and offset from the body. It is generally low and flat and has two distinct annules. P. brachyurus has a short strong stylet about 20 um long and large round basal knobs. It also has a short ventral overlap of the esophagus. The females have lateral fields marked by four incisures. Males are very rare and have one single outstretched testis.

Distribution and Economic Importance

Pratylenchus are second only behind Meloidogyne in terms of crops infested and the extent of crop damage and loss. Pratylenchus brachyurus is widely spread throughout the warmer regions of the world. In Brazil it is the most widely distributed root lesion nematode on coffee. They introduce necrosis in the cortex of the root which is detrimental to the plant. Lesion nematodes severely damage a plant and open points of entry for other bacteria and fungus to enter the root. In the US it is mainly found in southern regions and in Florida it is found in nearly every county. A study in 1969 showed that it is found in 90% of the citrus groves in Florida. Due to P. brachyurus being able to infest a wide range of crops and being so damaging, it is of high economic importance.

Reproduction and Life Cycle

These nematodes have the general nematode life cycle with four juvenile stages keeping their vermiform shape throughout the entire life cycle except the egg stage. Any of the stages, except the egg and J1 which molts inside the egg, can infect the root. The nematodes get inside the roots and move throughout the root laying eggs individually in the root cortex and maybe a few in the soil. This particular species of Pratylenchus reproduces by parthenogenesis. The males of this species have not been found very often.

Host Parasite Relationships

P. brachyurus can infest a wide range of crops. Peanuts, pineapple, soybeans, sugar cane, tomatoes, citrus, and cotton are just a few of the crops it can infest. With such a wide range of hosts, P. brachyurus is an important nematode for scientist to study and try to develop a resistant variety of its host plants. The problem with P. brachyurus is that that not too much is known about it and many of the reports available contradict each other. In 2001 in Nematropica, it was stated that the host range of these nematodes has not been determined meaning that crop rotation as a way to manage these pest could prove to be difficult. In 2012, the Journal of Cotton Science reported that this nematode is found in cotton fairly often but its pathogenicity is still in question. The variety of cotton used in the study seemed to be tolerant of the nematode and it only slightly affected the growth of the plant. Some varieties of cotton are intolerant to these nematodes so a farmer would want to try and plant tolerant varieties. When nematodes infest a root, they can open up the root to many other pathogens including bacteria and fungus. There have been numerous reports of nematodes interacting with Fusarium spp as well as many other fungi. Two scientists by the names of Michell and Powell presented evidence in their study that P. brachyurus can increase the incidence of Fusarium Wilt in cotton.

Management

Many different methods to control nematodes have been tried over the years. Some have proven more successful and others have not. Some of these methods include nematicides, cover crops, crop rotation, genetically modified plants and seed treatments. In soybeans, Essex is a tolerant cultivar whereas Forrest is sensitive. Essex has a higher yield in fields that are infested with the nematode.

The plots in this study by Koenning were treated with aldicarb, carbofuran, and fenamiphos and had less nematodes 40 days post planting then the areas that were not treated. The plots that were planted with Forrest which is the sensitive cultivar and treated with carbofuran had a higher yield than untreated control plots. Damage threshold levels have not been well established but using fumigated plots has shown significant increases in yield. In Brazil, many crops are planted under the no tillage planting system. This typically causes of buildup of polyphagous species of nematodes like P. brachyurus. If a no till system is being used, planting cover crops that would decrease the density of the nematodes may be needed. The problem with that is there is contradictory information regarding P. brachyurus and which plants are poor or good hosts. Due to this very same reason, farmers would have to be very careful in choosing a crop rotation schedule as well. In 1992 Vast demonstrated that arbuscular mycorrhizal fungi could enhanced the plant's resistance to nematodes by being a physical barrier. This would be something that could be used as a seed treatment to protect the newly emerging seedlings from being infested.

As with many other nematodes that cause crop loss, we need to continue to move forward in the study of these nematodes. Many of the managements tools used are very contradictory in studies and no good management strategy has been established. This nematode is so widespread and has such a wide range of hosts that several different management strategies will have to be developed based on the specific crop and area. Hopefully one day we will be able to develop a gene that can be put into plants to defend them against these nematodes as well as many other different types of nematodes.

Pratylenchus Loosi

Pratylenchus loosi is a plant pathogenic nematode infecting tea.

SHEATH NEMATODE

- Hemicriconemoides kanayaensis: Hemicriconemoides kanayaensis is a plant pathogenic nematode affecting the tea plant.

SPIRAL NEMATODE

Helicotylenchus Dihystera

Helicotylenchus dihystera (Steiner's spiral nematode) is a plant pathogenic nematode. It is known to inhabit sugarcane, rice, potatoes, corn, peanut, millet, sorghum, banana and forest trees.

STUNT NEMATODE

Tylenchorhynchus

Tylenchorhynchus is a genus of nematodes including many species of plant parasites. The classification of stunt nematodes - those including the Tylenchorhynchus genus - is unstable; many newly discovered species within this genus are reconsidered to be actually subspecies.

Stunt nematodes such as Tylenchorhynchus and the closely related genera, Anguillulina and Merlinia, include more than 250 known species. Members of these genera possess similar anatomy and may be easily mistaken for one another. Some debate has led to the classification of single species under different names in two distinct genera (e.g. Tylenchorhynchus cylindricus is Anguillulina dubia).

Tylenchorhynchus are soil dwelling stunt nematodes. They inhabit the same soil as plant root systems in which they can cause stressing or disease in plants. About 8% of the studied species are parasitic. Agricultural problems associated with Tylenchorhynchus spp. affect many species such as soybean, tobacco, tea, oat, alfalfa, sweet potato, sorghum, rose, lettuce, grape, elms, and citrus.

LEPIDOPTERA (BUTTERFLIES AND MOTH) PESTS

- Case-bearer species:
  - Coleophora scaleuta: Coleophora scaleuta is a moth of the Coleophoridae family. It is found in South Africa.
  - Coleophora euryaula: Coleophora euryaula is a moth of the Coleophoridae family. It is found in Egypt. The larvae feed on the leaves and fruits of Lycium europaeum and possibly Astragalus forskahlei.
- Ghost moth species:
  - Endoclita malabaricus: Endoclita malabaricus is a species of moth of the family Hepialidae. It is known from India. Food plants for this species include Acacia, Ailanthus, Albizia, Bridelia, Cajanus, Callicarpa, Camellia, Cassia, Casuarina, Clerodendrum, Coffea, Cordia, Eucalyptus, Eugenia, Filicium, Gliricidia, Gmelina, Grewia, Gyrocarpus, Herissantia, Lagerstroemia, Lantana, Macaranga, Mallotus, Ocimum, Rosa, Santalum, Sapindus, Solanum, Strobilanthes callosus, Tectona, Trema, and Ziziphus.
  - Endoclita punctimargo: Endoclita punctimargo is a species of moth of the family Hepialidae. It is known from Sikkim.
  - Endoclita purpurescens: Endoclita purpurescens is a species of moth of the family Hepialidae. It is known from Sri Lanka. Food plants for this species include Camellia and Cinchona.
  - Endoclita sericeus: Endoclita sericeus is a species of moth of the family Hepialidae. It is known from Java, Indonesia. Food plants for this species include Albizia, Camellia, Cinchona, Crotalaria, Manihot, Tectona, and Theobroma.

8

HRM Structures and Practices in Tea Plantation

PLANTATIONS

Plantations refer to large-scale agricultural enterprises, mostly specialized in one particular crop, grown for distant markets, employing a large number of wage labourers working under the close supervision in a strict hierarchical organization. Specifically, a plantation can be defined as an establishment for production of cash crops for export purposes. ILO defines plantations as an agricultural undertaking, employing hired workers, situated in the tropical or subtropical regions which are mainly concerned with the cultivation or production for commercial purposes of rubber, coffee, tea, sugarcane, bananas, cocoa, coconuts, groundnuts, cotton, tobacco, citrus, cinchona etc. It does not include family or small-scale holdings producing for local consumption and not regularly employing hired workers. In terms of global coverage ILO recognizes 15 commodities as belonging to the plantation category. Uma Devi Classifies plantation crops in Kerala into two groups viz., traditional and modern plantation crops. Traditional crops consist of coconut, pepper, cashew etc. Modern plantation crops are tea, coffee, and rubber.A major difference between these two groups is that the former was produced by the natives while the latter was introduced by foreigners who settled here to cultivate these crops. Generally plantation production refers to a 'large centrally operated estate which is usually mono cultivated and is operated by hired workers."

CHARACTERISTICS OF PLANTATIONS

Unlike other agricultural sectors, the plantation sector has necessarily to continue with the same crop for a long period. Plantation crops have a long gestation period ranging from 4 to 7 years before returns flow in. Once the gestation is over, production may take place for decades, even up to a 100 years, as in the case of tea. This necessitates the maintenance of a permanent labour force and providing them with housing facilities, medical care, education and other welfare measures. The plantations have a perennial nature of operations, which impose long term commitment of land resources and financial investment. Another feature is that it is a labour intensive agro-industry with land as the largest resource input in the production process. Each plantation has self-contained units of production and processing also. Corporate structure of holdings with managerial and technical expertise is yet another specific feature of the plantation sector. In the Indian situation the plantation industry is mainly export oriented. Hence fluctuations in international price, supply and demand affect the industry's prospects. In view of all these factors plantations are akin to an industry rather than agriculture. The plantation locations are remote places not conducive to mechanization. Unlike other industrial sectors plantations have little scope for expansion area-wise as there are restrictive land laws.

Plantations are characterized by a closely demarcated territory, an isolated labour force which live on plantation land and a market abroad. Manufacturing, transport and marketing of the plantation crops are handled by the plantation companies themselves. They tend to develop their own economic, social, cultural and political characteristics. Plantation production has certain peculiarities in terms of its origin, the use of labour, the type of labour used, the crops produced, its effects on the native society and native economy. In peasant farming unlike the plantations, the crop raised is usually a substance crop-which forms the main food item of the peasant raising it. The peasant sells the surplus or at times part of his reserves to fulfil his cash requirements. Use of hired labour is not very characteristic of peasant farming. In commercial farming both hired labour and modern technology may be used but they differ from those of plantations.

PLANTATIONS IN INDIA

The plantation concept is restrictively conceived in India as being applicable to tea, coffee, rubber and cardamom. Among these cardamom is the oldest and of indigenous origin. References to the medical properties of cardamom are found in Ayurveda treatment as far back as 1000 B.C. Next in historical reference comes coffee in the 17century. The plantation of Tea, Rubber and Cardamom has been declared as such under Constitutional provisions following which the commodity Acts of Coffee, Rubber Tea and Cardamom (spices) have been enacted to promote their development.This has been done because of the importance of these crops as major foreign exchange earners and of they being national industries spread over different states. In India the major plantation crops were indigo in Bihar, tea in Assam Nilgiri and Wynad, jute in West Bengal, cotton in Maharashtra, coffee in Karnataka, Tamil Nadu and Kerala, rubber in Kerala and tobacco in Andhra Pradesh as late as the 19and early 20centuries. Then came other modern plantation crops. At one time or another or in one region or another sugar, cardamom, cotton and pepper were produced in plantations.

PLANTATION SYSTEM

Plantation system as opposed to plantation production refers to the totality of institutional arrangements surrounding the production and marketing of plantation crops.It was in the sixteenth century that the plantation form of production emerged in most regions of the Third World. Uma Devi says that crops were usually introduced by the foreigners who had an imperialistic relation with the natives of the region and the plantation institution best suited the metropolitan needs in colonies of exploitation.The increased demand for Tea by the British consumers and the strained trade relations between China and Britain were two major reasons for Britain's pioneering efforts in the tea industry of India.China, the home of tea, was the sole supplier of tea to the European market. The East India Company was the agent for this trade. But the trade relation between China and Britain weakened during the period of 1780. This compelled Britain to turn their attention to India, one of their colonies for cultivation of tea on a commercial scale.

It was the British residents who played the leading role in re-organizing export oriented agriculture in the first half of the 19century.They were responsible for the creation of a new, imperialist order reshaping state and society in such a way that the individual agriculturist, preferably the British, could begin to cultivate those crops which were highly profitable in the markets controlled by British trading firms. The transfer of political power to British East India Companies marked the beginning of the new agricultural policy. A plantation sector emerged in South India in the second half of the nineteenth century. The European plantation sector (predominantly British) emerged here setting the political agenda to ensure growth and profitability.They demanded regulations intended to facilitate acquisition of fertile soil, recruitment of workers and export of estate products.The development of an estate sector was largely a phenomenon of the late 1850's and early 1860's.

In the early years P.D Devasahayam and some European planters started plantation production in South Travancore. In 1862 Daniel Munro opened Hope Estate in Peermedu range of hills. In 1877 J.D. Munro and others started plantations in Kannan Devan Hills. Gradually there emerged a planter's lobby in the 70's, 80's and 90's of the 19century. At first coffee was the only crop to register steady growth in this state. However in the late 1860's coffee cultivation was threatened by steam borers and in 1870's by leaf diseases.This resulted in the deterioration of vast numbers of coffee plants. There was also a decline in the price of coffee caused by the introduction of relatively cheap Brazilian coffee into the world market. Therefore during the late 1870's the planters began to consider alternatives. Initially the cultivation of cinchona appeared profitable. By the early 1890's tea became the most important plantation crop. Tea proved the most suitable crop for the Peermedu Vandiperiyar elevation. Tea took over as the most prominent crop in Kannan Devan Hills, too.

PLANTATION LABOUR COMPARED TO OTHER AGRICULTURAL LABOUR

Unlike in other industries, plantation labour requires open air work and constitutes a special kind of agricultural employment. It is for this reason that women are employed in large numbers in the estates rather than in factories. The work is unskilled and women from villages are far more used to this form of employment than working as skilled hands in other industries. A peculiar feature of plantation labour, especially in tea, coffee and cardamom plantations, is that it is recruited on a family basis from distant villages as most of the estates are situated in remote places of high attitudes where workers are not easily available. As a settled labour force is essential, whole families are usually employed. As wages are lower than in other agricultural jobs, every member of the family seeks work for economic reasons. Women are also paid less than men and are therefore employed in large numbers. A distinguishing feature of plantations workers, compared to other agricultural workers, is that they earn a regular income to support the family. They are among the most organized labour force of the economy. Moreover, the mechanization of plantation work does not pause a threat of unemployment to these women workers, as the scope of mechanization is limited in this area. The socio-economic background of plantation workers also differs from that of their counterparts in other crops. They come from different social, racial and cultural groups and are keen on preserving their traditions.

RECRUITMENT SYSTEM

During the pre-independent period, recruitment in the plantations of South India was through intermediaries called Kankanis,who work regular employees in the estates. Because of their wide contacts or ability to work or get work done, they were engaged to bring labour for which they received a commission from the workers they brought. A distinct feature of recruitment was that the unit of recruitment was the family and not the individual. Though the system of recruitment through Kankanis was abolished during the post-independent period, it still continues in certain estates. Many estates have settled labour of a generation or more, and on that account, local labour is available for recruitment. A large number of workers are employed as casual labour or through contractors. Trade unions and their leaders have a say in recruiting new labourers. Casual labourers are given preference. In the early days of plantation development, the only means of getting about was on foot or horseback or with bullock carts. The planters faced enormous difficulty in moving up with the newly recruited workforce from the plains. They had to provide shelter for themselves and their workforce. Hence the provision of accommodation for imported workforce became a necessity at the inception of the plantation industry. Facilities were also required for workers' health care, education of their children and supply of food and clothes. Housing, medical aid, water supply, and sanitation facilities were to be set up.

EMIGRANT LABOUR

As most plantations are situated in secluded uninhabited hilly areas, workers had to be recruited from far away places. Even as early as 1865, four fifths of the coolies in estates in the Nilgiris came from Mysore. According to the Report of the South Indian Planters Inquiry Committee of 1986, in 1877 thirty thousand Mysore coolies were employed in the Malabar-Wyanad area.When supply of labour from Mysore ceased planters recruited people form the densely populated Madras districts of Salem, Madura and Thinnnelveli. At the time of the Inquiry it was observed that south India plantations depended on an annual immigration for four-sevenths of their labour force.The situation was aggravated in 1895 when around 1,24,000 workers from south India went to Ceylon to work on the Plantations there.The planters had to compete with their counterparts in Malaya and Sri Lanka for labourers.In North Eastern regions of Assam and Bengal the planters imported labour from Chotta Nagapur area in Bihar and from villages mostly inhabited by tribals as well as the tribal belt of Madhya Pradesh and Orissa. In most cases families were brought and settled in tea plantations.

The pioneering planters went themselves in person to nearby villages and made forcible recruitment. People were reluctant to leave their family for long periods and hence family members were also taken along with workers. Thus family employment developed as a natural corollary of such a system of recruitment. The phenomenon of migrant labour is a common feature of the development of plantations all over the world. Land most suited for plantations are usually uninhabited hill tracts or densely wooded areas. Hence planters had to depend on migrant labour.

CONTRACT LABOUR

The shortage of labourers and the competition for labourers from Ceylon gave rise to the Kangani system of recruitment. The kanganies were given a cash advance in return for which they undertook to recruit an agreed number of workers. A maistrie's gang usually involved 20 to 40 workers. The maisteries in turn advanced money to the workers for their livelihood and for their family members and dependents. Later, the system proved to be detrimental to the workers' well-being as the kanganies were only interested in increased rates of commission; so the Government of India abolished the system. In course of time labour became available locally as more and more people joined the work force from migrant families who had decided not to return home and had settled down in planting districts after expiration of their contracts. By mid-twentieth century, the plantations witnessed a period of surplus labour; the industry was unable to offer work to the children of the workers but nevertheless had to assume economic responsibility for the increasing number of dependents.

SEASONAL WORKERS

In India, approximately 20-25% of the workers on plantations work only seasonally. This is an inherent feature of most plantation agriculture in India. In North East India, 200 million kg. of tea is produced in the 3 months from July to September accounting for 40% of the annual output. During winter Tea bushes do not produce any pluckable shoot. Hence arises the need for employing temporary workers during the boom period. The dependents of the permanent workforce resident on the estate are usually recruited whenever the need arises on a regular basis. In course of time they became absorbed into the permanent category. When the dependents of the permanent workforce could not fill all vacancies, unskilled and unemployed labour residing outside the plantations or people residing in the neighbourhood of the plantations were recruited. Lack of opportunities of employment in the vicinity of the plantations created in large number of unemployed job seekers in the families of plantation workers. These people served as seasonal workers.

WOMEN WORKERS

Plantation offers maximum employment opportunities to unskilled women labour. Of the total workers 56% are women employed in Tea plantations. Their services are mostly used for harvesting which is an exacting and delicate operation. Women are preferred for their manual dexterity and the care they take in plucking. Women in plantations often take primary economic responsibility to support their families. In certain cases women workers could earn more wages than men could because they were paid on piece-rate for plucking. The role conflict of woman as wife, mother and worker is applicable to the plantation sector, also.

Usually these women are self managed due to economic independence obtained through their direct participation in Tea production. Yet the chances of their upward mobility are nil. No women are appointed to supervising posts though TATA Company has appointed a few women as Assistant Managers. The hierarchical and patriarchal structure created through recruitment and organization of labour has in return promoted a semi-feudal attitude towards women.

Therefore women are considered to be inferior to men. The prevailing gender prejudice and the role conflict mentioned above have tied women down as ordinary workers. Moreover, women begin their work at 8 am and stop at 4.30 pm whereas men stopped their work by 2 pm. Plucking is paid on peace-rate while other field work is paid on time rate though a minimum quantity of work is fixed for both.

UNIONISATION

Griffiths observes that trade union movements in Tea plantations really began with the formation of the Estate Staff Association of South India in 1929.Technically it may not be considered a union as it was registered under the Societies Act. In 1947 a new association viz. Estates Staff Union of South India registered as a trade union was established and the majority of staff members have been well-organised under this union. The terms and conditions of service are negotiated and long-term settlement reached for the entire South India through collective bargaining between UPASI and its affiliate state level association on the one hand and the Estate Staff Union of South India on the other hand. Labour unions in South India Tea districts are of comparatively recent origin. It was in 1947 that the Waynad Estate workers Union submitted a written demand for this purpose. A number of other unions came into being about this time and in the course of the next few years unions multiplied rapidly.In 1961 there were about fifty five unions working in the tea plantation sector.In the beginning trade unions did make positive contribution towards improvement of workers' lives. The collective bargaining system helped to increase the wage level of workers. It created an awareness among workers regarding their rights and the need to fight for their rights. However, the political affiliations of unions and the petty interests of union leaders in fact, made the unions ineffective in improving the standard of living of workers. In many cases the trade unions failed to force the employers to implement the provisions of Plantation Labour Act. As tea plantation workers are by and large isolated they are not aware of the legal protection they are entitled to.

LEGISLATIVE ENACTMENT ON LABOUR WELFARE IN PLANTATIONS

Before independence, the only specific legislative enactment covering plantations was the Tea District Emigrant Labour Act, 1932. It was intended to regulate employment of labour in the tea gardens of Assam. Though Workman's Compensation Act, 1923, was applied to plantation workers, it did not confer any substantial benefit on them as accidents were recurrent in plantations. The Payment of Wages Act, 1936, though it applied to plantations, was concerned with the mode of payment of wages only. The Labour Investigation Committee (Rege Committee) noted that conditions of life and employment on plantations were different from those in other industries and recommended the creation of a Plantation Labour Code (PLC). The Plantations Labour Act (PLA) 1951 was enacted on the recommendations of the Committee.

THE PLANTATIONS LABOUR ACT, 1951

The living and working conditions of plantation workers are statutorily governed by the Plantations Labour Act, 1951, as also a host of other labour enactments. This Act is a Central legislation administered by State Governments through rules framed by them. The Plantations Labour Act covers both conditions of work and living in plantations. Provisions in the Plantations Labour Act (to be provided free of cost) are the following:

Conditions of work	Conditions of living
Duration of work	Housing
Weekly rest days	Medical care
Employment of women and children	Crèche
Annual and sickness leave	Canteen
Overtime	Education of children
	Protective clothing
	Conservancy
	Drinking water
	Recreational facilities

Although the original Act applies only to tea, coffee, rubber and cinchona plantations, the State Government has extended it to cardamom, cocoa and oil palm. The pattern of coverage outlined in Table 4.1 reflects the size distribution of plantations in each crop.

Table. Area Covered Under the Plantations Labour Act in Kerala (1999-2000).

Sl. No.	Crop	Total Area (Hectare)	Area covered under Act (Hectare)	Percentage
1	Tea	34,656	30,501	88
2	Coffee	82,348	8,070	9.8
3	Rubber	4,43,300	93,093	21
4	Cardamom	43,320	28,312	65

OTHER LEGISLATIONS

In addition to the Plantations Labour Act, 1951, conditions of work are covered by a number of other legislations. They can be divided into five categories.

- Wage-related
  - The Minimum Wages Act, 1948.
  - The Payment of Wages Act, 1936.
  - The Equal Remuneration Act, 1976.
  - The Payment of Bonus Act, 1965.
- Job security-related
  - The Industrial Disputes Act, 1967.
  - The Industrial Employment (Standing Orders) Act, 1946.
- 3 Employment, injury or death-related
  - The Workmen's Compensation Act, 1923.

- Social- security-related
  - The Employee's Provident Funds and Miscellaneous Provisions Act, 1952.
  - The Payment of Gratuity Act, 1972.
- Miscellaneous
  - The Factories Act, 1948.
  - The Maternity Benefit Act, 1961.
  - The Kerala Industrial Establishment (National and Festival Holidays) Act, 1958.
  - The Kerala Payment of Subsistence Allowance Act, 1972.
  - The Kerala Labour Welfare Fund Act, 1965.

Thus, a host of statutory legislations besides the specific Plantations Labour Act, regulate conditions of work as well as life of plantation workers.

OTHER STATUTORY BENEFITS

Apart from the provisions of Plantations Labour Act, all labour enactments applicable to industrial workers are also applicable to plantation labour force.

- Maternity Benefits: The Royal Commission on labour had recommended that maternity benefits should be provided to women workers in plantation by legislation. Women workers are entitled to maternity benefit for a total of Twelve weeks' wages.
- Provident Fund: Employer and worker have to contribute at the rate of 10% of the wages.
- Gratuity : Employer has to pay 15 days' wages for every completed year of service subject to a ceiling of Rs.50,000/- at the rate of last drawn wages.
- Bonus: Minimum bonus is 5.33. Maximum is 20%

THE INDIAN TEA PLANTATIONS

Tradition ascribes the discovery of Tea to Chinese people. The word tea is derived from T'e- in Amoy language. The use of tea was known to the Chinese as early as 2737 BC.It was used as a medicine in China in the 4Century. A Chinese scholar of eighth century, Lu Yu, describes the beneficial effect of tea as follows: "The effect of tea is cooling. When feeling hot, thirsty, depressed, suffering from headache, eye ache, pain in the joints, one should drink tea only, four or five times."Dutch trade introduced tea to Europe in 1660.It became the popular drink in England in 1664.

By 1700 tea had become one of the great non-alcoholic drinks in many countries. The Tea plant belongs to the botanical family of Camelliaceae. There are 30 genera and 500 species in it. It belongs to the genus Camellia of which there are 82 species which are mostly indigenous to the high lands of South-East India.Tea is the most important of all camellia spp. both commercially and taxonomically. The camellia was named in honour of a Moravian Jesuit called 'Came' (Camellus in Latin) who lived from 1660 to 1706 and wrote about plants in Asia.Tea plant is generally placed in the polymorphic species 'Camellia sinensis'. Recent findings show that this plant of commerce in India is derived from two species viz.

- Small leaved China variety (Camellia sinensis)
- Large leaved Assam variety (Camellia assamica)

However it is observed that more than one or two species are involved in the evolution of the present day tea plant of commerce in India. Considerable interspecific hybridization has taken place in nature.Hence the taxonomy of tea is compounded. However Tea today is botanically referred to as Camellia sinensis (L) O. Kuntze irrespective of species - specific differences.The tea plant camellia sinensis is indigenous to Assam and it was discovered there in the 1820's.

In 1823 Major Robert Bruce, who had been posted to the province of Assam, was told of the existence of tea plants in north-east Assam adjacent to Burma.Authors differ on the process of development of tea plantation in North India.The development of Tea Industry in South India actually started with the experimentation of Dr. Christie in Nilgiris in 1832.Commercial planting started in the South Indian region in 1853.In central Travancore tea planting started in 1864 and by 1906 there were 8000 acres of tea around Peermedu.At first tea was very much a subsidiary crop and in many areas it was started when coffee plantations declined because of a disease affecting leaves which appeared in the1870's. In the Kannan Devan Hills, large scale planting started in the 1880's and by 1896 six thousand acres were under plantation crop.In the Waynad area tea planting started in 1986 as an alternative or supplementary crop.

DEVELOPMENT OVER THE YEARS

The credit of establishing tea plantations in India largely goes to Britain. North east India was the pioneering area; where the government's first experiments in the cultivation of tea plant were conducted with both Chinese and Assam indigenous plants.As superintendent of Tea culture Bruce established nurseries and manufactured tea with the help of Chinese tea makers. In the early years of jungle clearance and planting the first generation planters had to undergo many hardships living in primitive mud huts and fighting with wild animals and diseases. The government, having ascertained that cultivation of tea plant in India was a feasible proposition, started handing over all its experimental tea lands to private enterprise.

Early in 1840 two thirds of the East India Company's tea lands were handed over to the newly formed Assam company rent-free for a ten-year period. The Assam Tea Company also took out leases on various other wild tea tracts.This can be said to be the beginning of tea plantation system in India. During the early 1850's there were many new arrivals and private planters acquired land on their own account and started tea plantation. The present day Assam companies are part of the earliest plantations which spread to the Brahmaputra valley, Cachar in Assam, Doors region of west Bengal and finally to Nilgiris, Anamalai, Munnar and Peermedu in South India. After independence many tea gardens changed hands and today most are Indian owned. Below is given the present tea plantation profile in India.

Table. Tea Plantation Profile in India.

Area (hect.) 1999-2000	437,857
Production (tonnes) 1999-2000	798900
Value of output (Rs in crores)	5820
Average exports (tonnes)	190200
% in productions)	24%
Domestic consumption(tonnes)	608725
Labour employed (lakhs)	11.38
Contribution towards taxes (Rs. in million)	2,500
Average productivity(Kg/Hect)	1,700
Normal gestation period (years)	3-4
Normal economic life (years)	60
No. of growers	54,000

HUMAN RESOURCE STRUCTURES IN TEA PLANTATIONS

Plantations are the largest single employer in the private sector of labour in the country. Among all plantations tea is the most labour intensive industry and the largest employer of organized labour. The peculiar nature of the industry, round the year harvesting etc. necessitate this. In tea industry labour cost forms about 50% of the total cost of production.The most important job on a tea estate is plucking of tea leaves.

The harvesting of tea leaves from the bushes in plantation goes by the name of plucking. This is done by carefully snapping off the tender shoot between the thumb and first finger. Basically, the harvest is of a bud and the two tender leaves below it because this is where the highest concentration of flavour producing chemical is. The timing of plucking operation has therefore to be geared to the rate of growth of the new shoots, which if course varies from season to season.

Plucking, although an apparently simple job, requires a certain degree of skill and great attention. Two-thirds of the total number of labour-days in tea plantation are ordinarily devoted to plucking. Women are usually employed in plucking tea leaves, because they are considered cleverer than men in this job. In fact, the plucking of tea leaves is the exclusive occupation of women because a skilled leaf picker is usually a woman, needing little supervision in her work. More over, some facts stand out which favours the employment of women. Women are obviously more steady workers and they are also more regular.

The maternal instinct also seems to be exploited. For the sake of keeping their children fed, they are less likely to risk their jobs by absenteeism. Again by making the wife a permanent worker, her husband is kept permanently available for work as and when there is demand. The children can also be drawn upon for cheap labour. Employees in tea plantations can be classified as below:

WORKERS

The term 'workmen' covers all employees in an estate, male and female except persons who are members of the supervisory, medical and welfare establishment or clerks or motor vehicle drivers who are employed on a monthly rates of pay.Both skilled and unskilled workers are included in the term. According to standing orders of March 1951 there are only two classes of workmen.

- Permanent worker. A permanent worker is one who has been passed by the employer as fit for work and has been registered on the check roll for the period of contract.
- Casual workers. Casual workers are those who are engaged for work of a purely casual or temporary character .

  The permanent or casual workers may be classified into three categories.
- Field workers: These people work in the estate as pruners , weeding and spraying workers or workers employed in the houses of managers, assistant managers etc. They are usually paid on time rate.
- Pluckers: These are people plucking tender leaves. Usually women are employed for this type of work. It is often a piece-rated work.
- Factory workers. Both skilled and unskilled workers working in the tea factory are included in this category. The wage rate is slightly different from the field wage rate.

STAFF

The term staff refers to members of the supervisory, medical and welfare establishment or clerks or motor vehicle drivers employed on monthly rates of pay and are not covered by individual contracts.The staff work under the managerial personnel at administrative and supervisory levels at different departments of a plantation such as office , field, factory, medical etc. The staff members are classified as permanent, probationary and temporary. There is another category of people especially with regard to wage settlement viz. special category.

Mechanics, electricians, wiremen, laboratory technicians, fitters, drivers, carpenters, masons, blacksmiths, midwives, balsevikas, teachers in unaided schools, medical orderlies, plumbers, pump operators and others doing similar work come under this category.

MANAGERS AND ASSISTANT MANAGERS

Managers and Assistant Managers are generally classified under the management cadre. Their terms and conditions of work are determined through individual contracts. In many North Indian plantations human resources are categorised into four strata viz. the management, staff, sub-staff and workers. The management comprises the manager, assistant manager and factory supervisor. The staff consists of white-collar workers. The sub-staff consists of lower level supervisors and other monthly-rated workers. All others are included in the category of workers.

PLANTATIONS IN KERALA'S ECONOMY

Plantations and homestead farms merge and mingle in Kerala. Plantations as a system differs not so much in the nature of crops grown as on the scale of operation, the systematic way the production is organised and carried on, the standardisation of the procedure and marketing proceedings and its access to the world market. Kerala is one of the most important plantation states in India. It accounts for nearly half of the total area under plantation crop in the country. Kerala has the unique distinction of being the only state having a substantial stake in all the four plantation crops coming under the plantation industry in the Indian context. viz. Tea, Rubber, Coffee and Cardamom. This is depicted in Table No. IV-.3

Table. Place of Kerala in India's Plantation Map.

Type of Crops	Area (in hectares)		Production (in tonnes)	
	Kerala	India	Kerala	India
Tea (1999)	36,877 (8%)	4,37,857	67,209 (8%)	7,98,900
Rubber (1999-2000)	472900 (85%)	5,58,584	5,84,680 (94%)	6,22,000
Coffee (1999-2000)	84,139 (25%)	3,40,306	60,470 (21%)	2,92,000
Cardamom (1999-2000)	40,867 (56%)	72444	6550 (71%)	9290
All plantation crops	634783 (45%)	1,409,191	7,18,909 (42%)	1722,190

Plantations in Kerala account for 24% of the net cropped area. However this sector contributes 25% of the gross value of income from primary industries.It provides direct employment to over 4.5 lakh workers. It also covers a considerable number of persons under secondary employment. The four plantation crops together contribute about Rs.53.5 crore to the state exchequer by way of agricultural income tax, land tax, plantation tax and duties levied by the state.Plantation crops on an average account for about 8% of state's SDP. Cropwise, rubber accounts for 60% followed by tea 24% coffee 10% and cardamom 6% in the SDP emanating from the plantation sector.All plantation crops except rubber earn sizable foreign exchange. Moreover there is hardly any import content in the production of plantation crops. This is significant in the view of an import content of 42% in the exports of manufactured industrial products. Thus, the plantation crops' net foreign exchange earning is almost 100 per cent. The plantations pay taxes and duties to the central and state governments on land, production, income, sales etc. The Government of Kerala gets state-sales tax/purchase tax, agricultural income tax and plantation tax. On an average the plantation crops together contribute 8% of the state's revenue, that is, under Rs. 100 crore.Plantations are labour intensive industries. On an average plantations provide employment to 3,60,000 workers around the year. The four crops account for as much as 18% of the agricultural employment in the state.

KERALA TEA INDUSTRY

Tea industry has a greater role to play in the state's economy than other plantation crops. As Tea is cultivated in the hilly tracts of the state, it prevents soil erosion of these areas and maintains ecological balance. Promotion of the industry enables government to maintain regional balance in development planning. The development of Peermedu, Munnar, Vagamon and other places in the eastern part of Kerala owes much to Tea industry. As tea is a highly labour intensive enterprise it provides better employment potential. Tea alone gives direct employment on an average to 83,000 workers in auxiliary units. In view of the fact that Kerala is facing acute unemployment, the importance of this industry should be obvious. It is only in tea that 50% of the cost of production is related to labour wages and benefits. Another distinguishing feature is that urban concentration and environment pollution are non-existent in Tea Industry.

TEA PRODUCTION

Kerala accounts for 36% of the Tea output in South India. Table No. IV-4 explains this.

Table. Tea Production (Tonnes).

Year	All India	South India	Kerala
1961	3,54,394	81,092	39,545 (48.76%)
1971	4,35,468	1,03,137	42.729 (41.43%)
1981	5,60,427	1,22,637	47.631 (38.84%)
1991	7,41,719	1,85,046	67.110 (36.20%)
1997	811m.k.g	47029	

Figures in brackets indicate Kerala's share in South Indian production. The per capita tea consumption in the state is 1.30 kg. per annum, the highest in the country. Yet 40% of the state's tea production is marketed outside the country. A total of 35,000 hectors is under tea cultivation in the state. It accounts for 47% of the area under tea in South India. 67% of the area is in Idukki district and 16% in Waynad. Kerala accounts for 45% of the total area of plantation crops in the country and 25% of the production from all plantation crops.

Table. Area Under Plantation Crops Vis-a-vis Total Cultivated Area in Kerala Area in '000 Hectares.

Year	Rubber	Coffee	Tea	Carda-mom	Total for plantations in Kerala	Total cultivated area in Kerala
1985-86	328 (!5.0)	66 (2.8)	61 (2.8)	35 (1.6)	490 (2.4)	2191
86-87	343 (15.5)	66 (3.0)	61 (2.8)	35 (1.6)	504 (22.96)	2209
88-89	367 (16.5)	75 (3.4)	64 (2.9)	35 (1.6)	540 (24.4)	2213
89-90	377 (16.7)	75 (3.3)	44 (2.0)	35 (1.6)	531 (23.6)	2252
90-91	3874	75	44	35	538	

ORGANISATIONAL STRUCTURE IN THE EARLY DAYS

During the pioneering phase, plantations were often opened up, managed by the individual owner himself. Many of these planters were either missionaries, their sons, or government servants. They were doubtless and adventurous persevering men. Those who opened up an agro-enterprise in the midst of forest areas risked financial loss and isolation from family and fellow countrymen. They lacked even the most basic facilities, including shelter and health. They were working and living side by side with their labourers, while clearing the jungle. The social distance between the planter and the labourers was relatively small in this early period. Initially, individuals and family firms were engaged in tea production. But as expenses soared, limited companies with salaried managers came into existence. The passing of the Companies Act encouraged this. The first joint stock company appeared in Peermedu/ Vandiperiyar in the 1890's. Around this time James Finlay and Company floated the Kannan Devan Hill Produce Company in 1897.

PATTERN OF OWNERSHIP

In Kerala the Dewans and other administrators gave wholehearted support to the British pioneers as the government could raise more revenue from the land by taxation.In 1879 J.D Munro and the Turner brothers got leased land from Poonjar Rajah. They transferred this to north Travancore Land planting and agricultural society. Later this was ratified by the government of Travancore and the tax to be paid was ½ a British rupee per acre.The proprietary ownership did not last long. Lack of sufficient working capital for paying wages and commodity advances was one of the reasons for early disappearance of proprietary ownership. Competition from new areas induced technological changes to keep costs down. This involved substitution of capital for labour. Further, the individual planters were not able to withstand the recurring fluctuations in the price of tea in the world market. All these factors contributed to the growth of Corporate Capital in the tea plantations of South India.

EMERGENCE OF THE CORPORATE FIRM

The Companies Act of 1862 in Britain encouraged large undertakings in the corporate sector. Hence, old family firms tended to be replaced by limited liability companies run by salaried managers. These developments were reflected in the changing conditions of the tea plantation industry in South India where proprietary planters continued to be in the majority. This process accelerated the shift towards better financial resource mobilisation and more stable professional management of plantation enterprise.

SIZE OF ESTATES

An estate is a standard unit of production in tea plantations. An estate includes the place where tea is grown and the factory where it is processed. They can be broadly classified into three types : a) Small growers b) Medium and marginal estate c) Big estates.

- Small growers: Generally estates having less than 50 hectares and no separate factory of its own are regarded as small growers. Generally estate rules are not applicable to these estates. In Kerala there are a good many tea growers owing up to 5 hectares.
- Medium and Marginal Estates : Estates having less than 200 hectares, with or without a factory of their own are treated as medium estates. Their yield is much less compared to the big units because of poor cultivation techniques or infertile soil or bad weather so on. Yet, they are bound by all estate regulations.
- Big Estates: Estates having an area above 200 hectares are classified in this group. These are mostly owned by big companies. One main characteristic of these companies is that they are producers, exporters and retailers of tea in India and abroad. In South India this group accounts for 64.58% of the total area.

Economic Size

The size of an economic holding is determined by various agro-economic factors such as climate, elevation, soil availability and cost of labour, transport facilities, capital, management and equipment. Costs and yields vary very much according to geographical location. Hence an economic unit on the basis of acreage in one location may not be economic in another. Management also is a varying factor. The price factor may covert an economic unit into an uneconomic unit. The Plantation Inquiry Commission (1956) on tea points out that an estate of at least 300 acres would be required to feed a factory of the smallest size throughout the year. Larger Estates facilitate the adoption of improved cultivation techniques and thereby lead to increased productivity. A small estate cannot raise adequate green leaf to economically justify the construction and operation of a factory of the smallest size and has to depend on bought leaves. An analysis of size-cost relationship becomes imperative in this connection. The survey conducted by the Tea Board in this regard covering 135 single-estate tea companies in India over the years 1958-1962 revealed that estates between 3000 and 4000 hectares had the highest productivity per unit area as also the lowest unit costs of production.Neither the area nor the yield per unit represents a valid criterion to determine, whether an estate of a given size could achieve an adequate economic return. However, they are generally adopted as such for want of a better criterion. For the purpose of licensing new plantations, the Tea Board considers as uneconomic estates below 150 acres if owned by an individual and estates below 300 acres if owned by a company.

ESTATE MANAGEMENT

An average estate having 400 hectors will have a manager. The estate will be usually divided into 3 divisions. Each division may have one assistant manager depending on managerial compulsions. Under each Asst. Manager there will be 3 to 4 field officers. The field officer will have his own assistant field officer. Under the assistant field officers there will be 15 to 25 supervisors. Each supervisor will have 30 to 40 pluckers or 10-20 field workers under him. In some cases there will also be a visiting agent (V.A) under the general manager. He will be an experienced manager and will be acting as a super manager. The supervisor will be in charge of 20-30 workers in the case of plucking and 10-12 in the case of other workers.

ORIGIN AND GROWTH

Unlike other agricultural or industrial enterprises plantation management has evolved certain distinctive features. Management in tea plantation owes much to the British and other pioneers for its style and functioning. The pioneers developed the style from their experience and from the specific needs of the industry. The management system largely evolved before the emergence of organised labour and the enactment of different legislative measures. This was not in tune with the developments in management science, as we understand it today. In the recent past considerable research on management practices has been carried out and many new techniques of management have been developed for application in industrial organisations. But this has had little impact on the plantation industry. In the beginning the forms of organisation in tea industry were mainly of proprietary and partnership nature. Later joint stock companies came on the scene with large holdings of estates. These joint stock companies altered the form and style of management.

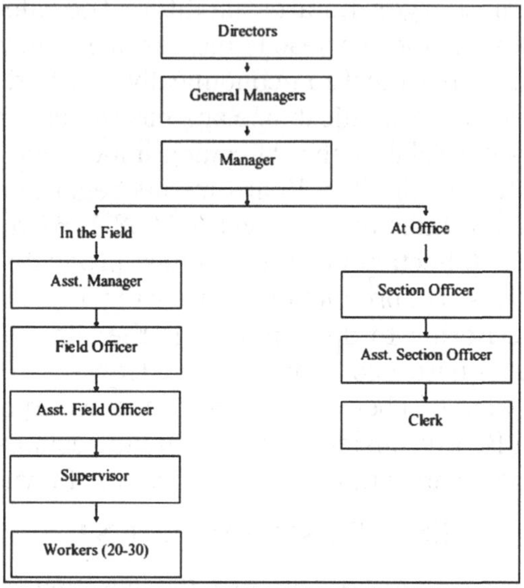

Fig. Organization Chart of an Estate.

ROLE OF MANAGING AGENCIES

Managing agency was a business firm developed to provide managerial, financial and technical services to estates whose owners were away from the properties. These agents provided services in budgetary control, cultivation and manufacturing advice, taxation advice, exchange control and staff recruitment. Many of the functions of the present estate manager were transferred to the managing agents. They were giving orders virtually to the managers. The managing agencies were organised very much in the form of a Company. As owners of small estates could not afford to have these services these estates were consolidated for their services and thus had access to the developed capital market of London. In the beginning the typical unit of the tea industry in South India took the shape of a Sterling Company in Britain owning tea plantation in South India. The agency houses, many of which were situated in Britain, started their own ventures and took over some of the sterling estates to look after the interests of Sterling Companies. In course of time, the early individual planters became associated with agency houses that managed several tea plantations, and thus incorporated into the firm structure of managing agency System. Thus the capital of managing agents and the managerial experience of the proprietary planters provided a sound foundation to the growth of tea industry in South India. Messers Harrison & Crosfield, James Finley, Brook Bond Ltd. Peirce Leslie etc. were some of the leading managing agents in the field till the system was abolished. The link between the agency firm and the estate manager was the visiting agent (V.A). The visiting agent was an experienced practising planter holding a fairly senior position. He used to visit the estate and give report to the managing agents. He was expected to give advice to the manager. Gradually he got more power and authority in the managing agency system. With the passing of the managing agency system the visiting agents' position was assumed by the Managing Director of the Companies. However, the functions of Managing Director were quite different from those of the V.A. He was more concerned with general Administration of the Company as in any other industrial concerns.

THE MANAGER

A manager in an estate is a full-time manager. He is manager for 24 hours. The peculiar situation in tea plantations warrants this. Since all the labourers are residential labourers the manager is responsible for every minute case in the estate. The Plantation Labour Act prescribes a lot of facilities like housing, electricity, drinking water, hospital etc. to be arranged for the workers. Very often a plantation is a closed community away from urban and rural contacts. Hence the workers are compelled to depend on the estate manager for a variety of services. This has led to a paternalistic style of management in some cases and a dictatorial style in some others. Unlike other manufacturing concerns, plantations generally employ more workers per unit of investment. The workers are disposed over a wide area ranging from 400 to 600 hectares. In the manufacturing industry the workforce is located generally on a small area. Hence supervision and control of workers is of especial significance in plantation industry. Levels of management and span of control are also distinct from those of other concerns. As a result, an efficient network of transport and communication and regular inspection is required in this area.

Manager, an Agriculturist

The management of an estate necessarily involves the management of agricultural process. The manager is expected to have a thorough knowledge of it right from breeding the plants, nurturing, manuring, weeding, and leaf collection, down to the factory process. Knowledge of different varieties of plants, their culture, the soil composition, better methods of harvesting, pest management, disease management, weed management, nursery management, shade for tea and its management a sine qua non of efficient management. Therefore, tea companies prefer people who have basic knowledge of the plant and agriculture for the managerial cadre. In the manufacturing field, on the other hand, the manager is more concerned with man and machinery. Hence, the method and practices in the field of manufacturing concerns cannot be applied with equal efficiency level in the plantation field. And so, while we maintain that management principles are of universal application both in business and in non-business concerns, it is important to take account of the quite peculiar work environment of the plantation industry. With this proviso it must now be considered whether the benefits of modern technology, method and practices in management functions could be applied to the plantation sector also.

Factory Manager

A profitable estate will have a factory of its own. The manager of the estate will be in charge of the factory, too. Unlike fieldwork this work will have to be done with more precision. Some kind of technical competence needs to be acquired in this line. In these days when computerisation is the order of the day, a tea factory needs also to be computerised. The manager should also be well acquainted with the characteristics of different varieties of black tea, their marketing opportunities, promotion methods etc. The workers in the factory are treated as a separate group and the manager must know the various Acts and rules that govern their work and their service conditions.

PLANTERS' ASSOCIATION AND MANAGEMENT

There are powerful employers association in the tea industry which exert great influence on policy making, wage negotiation and industrial legislations. One special feature of planters' association is that it was not formed as a counterpart of labour unions. The association had its origin way back in 1860 long before labour unions were organized in the plantations.However Ramachandran Nair argues that associations of employers have acquired the character of a trade union as the majority of associations are registered under the Trade Union Act. Planters' associations took their start at the socialization get-togethers at the planter's houses. From these gatherings the district planters' associations emerged as they were occasions for discussion on planting business. All planters, irrespective of the nature of their crops, take out membership in the association. The whole of Kerala is divided into different planting districts depending on geographical features and each district has an association of its own. Major district associations are

- The South Travancore Planters' Association
- The Central Travancore Planters' Association
- The Mundakayam Planters' Association
- The Kannan Devan Planters' Association
- The Nelliampathy Planters' Association

At the apex level in the state there is Association of Planters of Kerala (APK) and for South India as a whole there is United Planters Association of South India (UPASI) At all India level, there is the Tea Board under the Ministry of Commerce involved in target setting for production, exports, regulating the auction system, providing various incentives and imports for the industry. Below the Tea Board there are various other associations. The oldest and biggest among them are Indian Tea Association (ITA) and Tea Association of India (TAI). These two largely represent the North Indian Tea growers. South India is represented by UPASI which, apart from tea, represents coffee, rubber and cardamom planters.

PARTICIPATIVE MANAGEMENT

Unlike in any other industry, planters association plays a very important role in the management issues of planters. The associations make representation to the government on various matters like taxation, land availability, wage fixation etc. They also deal with various unions in the fixation of wages on a state wide basis or in some cases on district basis. The government for their part consult the association on important matters and the representatives of the association are appointed to different committees and Boards. The association of planters of Kerala now deals with labour problems on association basis and not on estate basis. There are sub-committees to deal with labour problems, liaison work with Apex bodies and district associations. The association played an important role in resolving labour unrest in matters of scarcity of food, clothing and inadequacy of wages.The member estates were advised to raise the wages of workers in proportion to the cost of living. The association is a member of the tripartite labour conference (PLC) where the government and labour unions make up the other members. These labour conferences recommended formation of works committee in plantations. The PLC played major role in adopting the gratuity scheme, abolition of kangany system in 1962 and periodic wage settlement. The decisions made at APK are binding on all plantations in Kerala. The association also participates in discussions with the Estate Staff Union of South India (ESUSI) and the minimum wages advisory board of the government. Unlike workers, the terms and conditions of service are negotiated long term agreement reached through negotiations on a South India basis between UPASI and its affiliate state level association on the one had and ESUSI on the other. The wage boards simply incorporate these settlements as their recommendations. These settlements are reached with a true sense of give and take as the Unions have never had recourse to agitation or third party intervention to redress their grievances. Differences of opinion arising from the implementation of the provisions of settlements are resolved at bipartite level.

9

Entrepreneurship in Small Tea Plantation: A Case Study of Assam

The typical plantation economies an outcome of Colonial era could be witnessed in different sectors and across the globe. In Brazil and Malaysia it was rubber, in West Indies it was sugar and in India it is tea. The tea industry of India is one of the oldest and perhaps the most efficiently organized agricultural enterprises in the country. The credit for the growth of tea industry goes to the British entrepreneurs who invested large capital during the initial days of plantation. After the Yandabo Treaty (1826), British had acquired full control over the entire territory of Assam; which has given incentives to the British traders to expand their area under tea plantation by acquiring Government land at a very favourable term.

After losing legal monopoly of the tea trade between China and Britain (1833), the East India Company began to look for other sources of supply in the newly conquered regions. Assam was selected as the most accepted region for tea cultivation because in the early 19th century tea was discovered growing wild at many places of Brahmaputra valley. The first experimental tea cultivation in Assam was started in 1836. In 1839, and first made tea was sent to London Auction and received a very positive feedback. This important development was the starting point of the tea industry of Assam. In tea plantation, the regulations enacted by the then colonial rulers were extremely favourable to the European entrepreneurs.

Due to lack of capital and knowledge over tea cultivation, local entrepreneurs have taken little initiatives in tea cultivation during the initial days. Even after the independence, the situation has not change much as British companies were taken over by the private parties/companies who belong to main land of India. Therefore it is rightly said that "the plantation sector in Assam thieved on alien interest, employed capital, labours from outside the region and catered to the needs of external market" (Nath,1997). From the modest beginning, Tea Industry of India at present has a total business of Rs.19000 crore and expected to reach Rs 33000 crore by the year 2015 (estimated by ASSOCHAM, 2011). It is the second largest producer in the world (988.32 mkg) after China and fourth in world export. Tea (chai) has become a common beverage for the country which consumes around 90 per cent of her total production. It is one of the oldest industries and among the largest employers in organised sector. Over 12 lakh permanent and almost same numbers of casual and seasonal workers are employed by the industry (Gothoskar, 2012). The growth of this industry in India was responsible for the development of infrastructure and the regional economy of many backward regions like- Assam. Tea is produced mainly North and South India. Assam dominates the countrys tea map by producing more than half (51.48 per cent) of total tea production. In recent year, tea industry of India has witnessed many structural changes, which includes emergence of tea cultivation on small holding, rising competition in domestic market, introduction of Bought Leaf factories(BLFs), (Das,2009). Now, production of tea in India takes place in both large and small scale. Cultivation of tea on small holding of land is a new concept in India. According to Tea Board of India (TBI), Small Tea Growers (STGs) has been defined, as a person or group having plantation area up to 10.12 hectares .At present STGs are became a very important constituent of tea production as it produces nearly 24 per cent of the total tea production of the country and nearly 1.58 lakh small cultivators are directly engaged in tea cultivation.

Table. Tea Industry in India at a Glance (2011).

	Number of gardens	Area under production (in hectare)	Total production (million k.g)	Employment
Small Tea Garden	157504	1.62 lakh	260 mkg	1 million
Big Estates	1686	4.18 lakh	728 mkg	2 million

SMALL TEA PLANTATION

Traditionally, tea is best known as plantation crop. Tea plantation is often called tea estates/gardens. Plantation comprises a large area, producing a single agricultural product as a commercial venture. This model (plantation) was introduced in Indian tea industry during the colonial period by the British planters. Until 1950s, it was thought to be the only way to produce tea on commercial basis where production to sale was controlled by the tea planters themselves. Tea cultivation on small holding belongs to a farm or production unit that allocates its labour, land or other resources for subsistence or semi subsistence income from it. Thus, a small grower is the owner and worker as well (Ahammed, 2007). The growers produce only green leaves; manufacturing and distribution parts are managed by different persons or groups. Cultivation of tea on a small holding is a usual practice in many major tea producing countries like- China, Japan, Sri Lanka, Indonesia, etc. The concept of small tea cultivation came into existence when Kenya in 1950s, had decided to produce tea for export. The Kenyan successful experience had created a model of small tea growers in developing and underdeveloped countries. Since then, there has been a steady shift in tea cultivation from big plantation to small holdings (CDPA, 2008). Today, in most of the tea producing countries, tea cultivation on small holding significantly contributes to the countrys total tea production along with the large estates. Inspired by the contribution of STGs in leading tea producing countries, many developing countries like Vietnam, Bangladesh, Nepal, Pakistan, etc. are also utilising their unexplored vacant lands in the vicinity of traditional tea growing areas for plantation of tea on small holding.

SMALL TEA GROWERS (STGS) IN INDIA

Tea cultivation on small holding is a recent development in Indian tea sector. This concept was initiated in 1930s, beyond the traditional tracts of cultivation in Nilgiris, Tamil Nadu (Chai Time, 2007). It was only late 1980s or early 1990s, the cultivation spread to the other tea producing states of India like - Assam and West Bengal. After independence, during the successive five year plans the TBI has spent all its resources in trying to mobilise big tea estates/plantations to increase production or expand area under tea. Most of its efforts borne little result (Bhowmik, 1991). Even with the Government investment through TBI and other agencies, there has been a marginal increase in production. From the 8th Five year plan (1992-1997) TBI has laid down some emphasis on the promotion of STGs. It has decided on this scheme because all other attempts to increase area under tea have failed. Therefore, it is proposed to encourage landless labourers and unemployed youths to take up tea cultivation in the tea growing areas of India.

GROWTH OF STGS IN ASSAM

The tea industry of Assam is more than 180 years old, but cultivation of tea in small scale is a new development. During the initial days, tea cultivation was quite colonial in nature without the involvement of native people. Few initiatives were taken by the end of 19 th century by the local entrepreneurs. This was limited within the influential classes and not permissible to common people (Gogoi, 2011). The local people never got prime land for tea cultivation (Deka, 2008).

After positive feedback of London Auction (1839), the Government encouraged British entrepreneurs to invest in tea cultivation. The investors were given land on easy terms and the policy of Colonial rulers was to favour their own people. Laboureres also hired from outside the state like- Bihar, Orissa, Madhya Pradesh and other parts of the country. There was a dearth of indigenous labour and capital and the commodity market did not evolve within the region (Nath, 1997). Besides, tea cultivation was a new concept for the local people; they might have lack of knowledge and interest to take tea cultivation.

Having their own land, they did not want to work as a tea labourer in the gardens. This stigma remained with local people more than 150 years. The mistake they have committed during the initial days of tea cultivation deprived the whole indigenous community from the industry and became the silent observers of the growth and development of tea industry of Assam. After independence, certain restrictions were imposed on cultivation of tea to keep control on quantity and quality of tea produce in India. Under Plantation and Labour Act (PLA) 1951, restrictions were imposed on tenure system, land holding, land transfer, minimum wage, etc. and PLA-1953, which necessitates a prior approval from Tea Board of India for tea cultivation. Tea Board of India is responsible to regulate output, quality and marketing of tea.

It gives approval on cultivation of tea on virgin soil, subject to fulfillment of certain conditions such as proof of ownership of land and size of the holding on which plantation of tea is proposed, proof of suitability of soil for cultivation of tea, proof of financial soundness etc. All these restrictions stood as hurdles to the growth of tea cultivation on small holding. Traditionally agriculture sector of Assam was underdeveloped and subsistence in nature due to lack of irrigation and investment, small holding, difficult terrain, insufficient Government intervention. It compelled the agriculturists to look for alternative livelihood. Tea cultivation on small holding was one of the alternatives for the agriculturists to sustain their livelihood in the long run as cultivation of tea provides works and income throughout the year. This can be termed as the beginning of an Indian entrepreneurship in the tea sector (Deka, 2008). In the present context, in India there are limited scopes for the growth of big tea gardens as after independence a lot of changes have taken place in the Government legislations to improve the socio-economic framework of the country.

So, the alternative for the growth of tea industry in India is left only setting up of small tea gardens (Baruah, 2008).Finally, the Government of India as well as Assam Government allowed the local people to open small tea gardens by abolishing all the barriers relating to the growth of this sector (Saikia, 2008). The first effort to popularise tea cultivation in small holding in Assam was made in 1978 by Late Soneswar Bora (the then Agriculture Minister of Assam) 1. His intention was to utilise available fallow land and attract young generation to agriculture sector and thereby to solve unemployment problem (Baruah, 2011). The abundance of uplands, suitable climate, availability of agrotechnology services, skilled surplus labours from nearby big tea estates, ready market to sell the green leaves in the estates factories, the advantage of plantation crop over the seasonal agricultural crops were the factors that encouraged the small and marginal farmers who were engaged in other economic activities to take up tea plantation (Das, 2010). The initial success was phenomenal. At present there are 700002 (approx) STGs in Assam with a land coverage of 1, 18,000 acre; contributing nearly 30 per cent of the total tea production of the state. On the other hand, the acreage under big tea estates declined from 4,09,684 hectares to 4, 01,512 hectares (1.99 per cent) during 1991-2006. According to a survey conducted by the Government of Assam, the major concentration of STGs (94 per cent) is found in Upper Assam districts (Tinsukia, Dibrugarh, Sivsagar, Golaghat and Jorhat). It is observed that the growth rate of STGs over the years is phenomenal. It has grown from mere 16 small tea gardens in the year1978, increased to 68,465 in 2011. Figure shows the trend of green leaf production (mkg) from the year 2004-2009.

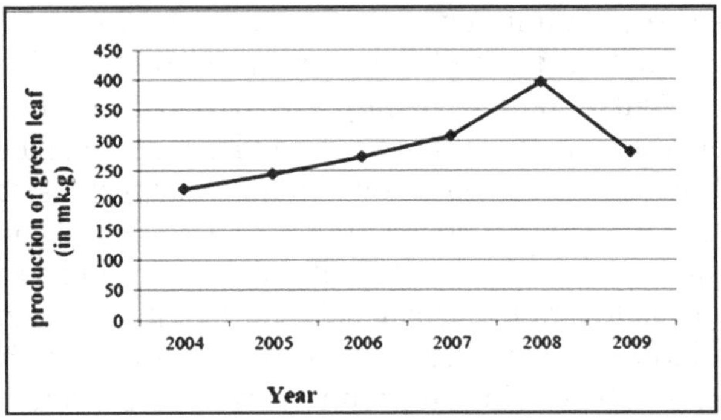

Fig. Growth of Green Leaf Production by STGs (in Mill k.g).

In the year 2004 production of green leaf was 218.41mkg and it rose to 396.574 mkg in the year 2009.

ENTREPRENEURSHIP IN SMALL TEA CULTIVATION

The economic importance of small business in a developing country like India is much more than innovative and invention; they have been a vital source of absorbing unaccountable no of person at place in semi rural areas (Maurya, 2001). It requires low capital and low level of barriers to entry. In India, besides unemployment the economy is facing many other problems like poverty, illiteracy, poor health, etc. To overcome some of the basic unsolved problems like unemployment and income generation, country may encourage entrepreneurship through small business. They are characterized by the relatively labour intensive choice of production, which contribute to the society by enhancing the employment capabilities of skilled and unskilled, educated and uneducated labour force which may otherwise remained unutilized. The opportunity to enter into this business venture may cultivate the spirit of entrepreneurship which will empower the economic growth of the society as well as the economy. Entrepreneur is an economic agent who plays an important role in the economic development of country. It is very old concept according to which, anyone who runs business is called an entrepreneur. Entrepreneurship is basically concerned with creating wealth through production of goods and services, which in turn helps in the growth of per capita income of a country. It speeds up the process of activating factors of production leading to a higher rate of economic growth, dispersal of economic activities and development of backward regions. Entrepreneur looks for innovative ideas and puts them into effect for economic development Therefore entrepreneurship occupies a crucial part of economic development by triggering the economic activities for the development of a region.

The relationship between entrepreneurship and economic development is very simple i.e. entrepreneurship creates new business; new businesses on the other hand create new jobs, intensify competition and may increase productivity. As a result, entrepreneurship will lead economic growth (Acs, 2006). Both empirically and theoretically it has been proved that entrepreneurs are the key force for the economic growth of a country (Karlsson, Friis, Paulson, 2004). Global Entrepreneurship Monitor (GEM) also supports this view and observed that there is no country in the world with higher level of entrepreneurship and low level of economic growth (UNCTAD, 2004). In order to know how entrepreneurship is good for development, it is necessary to aware the types of entrepreneurial activity in a region or a country is actually involved. According to GEM, Entrepreneurship may be divided into two:

- Necessity based
- Opportunity based

It is observed that in every country, there is some level of necessity and opportunity entrepreneurship. When a person becomes an entrepreneur because he has no better option is called necessity entrepreneurship. On the other hand, opportunity entrepreneurship is an active choice to start a new enterprise based on the perception that an unexploited or underexploited business opportunity exists. Opportunity entrepreneurship is one of the important factors for employment creation and acceleration of economic growth. It is possible that necessity entrepreneurship may not lead to economic development because being pushed into entrepreneurship (self employed) as all other option for work either absent or unsatisfactory can lead to underdevelopment. On the other hand opportunity entrepreneurship represents the voluntary nature of participation and therefore entrepreneurs expect their venture to produce more growth firms and provide more jobs (Acs, 2006). Therefore, it can be said that entrepreneurship plays an important role in the economic growth and development of a nation. It is purposeful activity which includes initiation, promotion and distribution of wealth and services. An entrepreneur is a crucial factor in economic development and an integral part of socioeconomic transformation.

It is a risk bearing activity and challenging task, needs utmost devotion, total commitment with fullest involvement for his personal growth and prosperity (Baruah, 2009). Entrepreneurs undertake business activities, employ themselves in that business and open up employment avenues for the others. Therefore, it plays s a fundamental role in a country like India. It is now widely accepted the fact that the active and enthusiastic entrepreneurs can explore the potential and available resources of country such as labour capital and technology. Entrepreneurship helps in the process of economic development through increase in income, mobilizing the available resources, creating wealth, generat ion of employment , etc. Cultivation of tea on small holding can be one of the most convenient sectors for developing entrepreneurship culture among the rural people of Assam. Tea growing in any scale, either big or small is an enterprise since it encompasses all elements of production viz.

land, labour, capital and organization (Taparia, 2003). Many studies (Mwaura and Muku, 2007; Amartunge and Shiratake,2002), has highlighted the fact that small tea enterprises are playing a major role in reduction of poverty, generation of employment etc., more effectively than their large counterparts in many developing and under developed countries. In Assam, the cultivation of tea in small scale is an outcome of the entrepreneurial ability of a group of local youth in the latter part of 1970s (Goswami, 2006). Active participation of the indigenous people since its inception provides sustainable livelihood opportunity to large number of rural population. It provides avenues for self employment to educated unemployed youths, generating additional sources of income to the farmers and engaging family members directly with this enterprise. Assam has been experiencing a social revolution in the field of tea cultivation on small holding for the last 30 years. It is an active choice of the farmers because, Assam is one of the finest places of producing tea in the world .Tea, as a commercial cash crop, has an increasing demand in and outside the country.

Natural factors like -soil, weather is highly favorable for tea cultivation in Assam. Infrastructural facilities like factories for tea processing, auction market for made tea, available land for tea cultivation, cheap labour in and around tea growing areas, technical and other necessary technical knowhow, incentives provided by the State for small tea cultivation are the biggest advantages available for the local entrepreneurs. These opportunities act as a big push-factor to many rural unemployed youths to take tea cultivation on small holdings. Tea cultivation on small holding is a family run enterprise. Family members are actively involved in maintaining the garden; only seasonal labours on temporary basis are engaged in these gardens. Small tea gardens are not coming under the domain of PLA, 1951 which gives an advantage to the STGs and thereby reduces the cost of production3. Initial Investment for tea cultivation on small holding is very less as investment requires only for plantation.

Crop failure in agricultural sector is a regular phenomenon in Assam due to lack of irrigation, flood, lack of technology in production, etc. But in tea cultivation, crop failure is limited very occasional pest attack and natural calamities (Banerjee, 2011). Tea cultivators have used suitable areas which were either un-utilised or underutilised. It is estimated that nearly 20 million hectare of land in upper Assam has been brought under tea cultivation by the small growers in the last decade. According to All Assam Small Tea Growers Association (AASTGA), at least five workers are employed in each small tea garden. It is estimated that about 2.40 lakh people are directly employed in this sector out of which 1.45 lakh is ex-tea garden labour or excess number of workers from the nearby big tea estates, specifically women pluckers. Engagement of tea cultivation by the STGs also contribute additional income to the total income .The average income of the STGs is Rs.49,598.264, which is much higher than the per capita income of the state (Rs. 27,197) at current price in 2009-10. With the increase in the production of green leaf in the state, manufacturing facilities like Bought Leaf Factories (BLFs) has been set up to help processing the green leaf in the tea cluster. There are 220 BLFs in Assam with average employment ranges between 20-25 workers and producing nearly 29 per cent of the total made tea production of the state.

Many new business opportunities and services like- supply of garden implements, transportation of green leaf and retailed outlets of agro chemical and manures, leaf agents, packaging, etc. Additional output of tea coming from STGs will help to meet the national and international demand for tea and also support in maintaining exportable surplus which will help in earning foreign exchange for the country. Average production of green leaf at STGs is 3500 k.g per hectare; which is much higher than the average productivity of the big estates (i.e. 1690 kg per hectare). Price of green leaf has also an important motivator in taking up tea cultivation in Assam. In the last decade the price of one k.g green leaf was sold in the range between Rs.8 to Rs.24 per k.g. It also gives positive attitude for utilising their available lands suitable for tea cultivation. Involvement of local youth in tea cultivation will create some role models in the rural area which will encourage the future growth of entrepreneurship in tea sector. Establishment of tea gardens in new areas will improve the local and regional environment. It will help to check soil erosion and destruction of natural vegetation since tea plantations would cover land.

MAJOR CONSTRAINTS

Tea small holdings are never expected to assume the mode of dominant producers. They are seen only as a means for providing some additional earnings to the peasantry, and thus absorbing of rural surplus labours (Das, 2012). In Assam tea cultivation on small holding is still in nascent stage. There are some hindrances for the growth and developments of this sector which include – ownership of land, lack of finance, low price realisation of green leaf, unorganised nature of cultivation, etc. Most of the growers are cultivating tea on Government ceiling land without having formal ownership. The Land Act of Assam restricts the transfer of ownership right to the growers. This restricts the growers to qualify for registration under Tea Board, unable to take the advantages of various schemes of Government, financial intermediaries. Whatever the growth has been taken place in small tea cultivation in Assam, the credit goes to the private initiatives of individual growers. Since, STGs do not posses their own processing unit; and have to sell their produce to the big estates or BLFs. being placed in the situation of dependent subcontractors, tea small holding remain in a stage of disadvantaged. Tea small holding have just become a convenient source for outsourcing by the estate sector, which set price for raw tea leaves. Sometimes these factories simply return raw leaves, citing inferior quality or compel small holders to accept the unremunerative price (Das, 2012). Therefore they are bound to sale their produce at whatever the prices, offered by the estates or the BLFs. the deceleration of the price of green leaf so rapid in recent years that the price offered by the large tea estates /BLFs has come down from Rs 21/k.g (in August) to Rs.4/k.g (in October),2011. BLFs enter into agreement for green leaf mostly with intermediaries like leaf agents or growers for volume of supply of green leaves. As the STGs are un-organised and un- unionised, they have the little bargaining powers with the factories. Lack of market information, absence of proper storage facilities, transportation problem, etc., ultimately bring down the price of green leaf. Moreover, the Government imposes cess like leaf cess (0.20 paisa/k.g), land cess (Rs.22/k.g) on the growers. Other than, this they have to pay a fine of Rs 200/ bighas if they put the tea under government fellow land. Unregulated growth of this sector in Assam has given raise the problem of price fluctuation of green leaf in the market. Without going through the marketing and economic holding etc. many small tea growers started cultivation of tea in the remote areas of Assam enthusiastically without technical knowledge; which ultimately leads to the uneven development of this sector.

POLICY RECOMMENDATION

- It is observed that small tea plantation may rebuild the socioeconomic structure of the rural indigenous people of Assam. The problems which are highlighted like land ownership problem; can be shorted out by the Government taking example from other states.5 The only thing needed here is the sincere effort from the Government side and look this small tea gardens a new development initiative for the state.
- If we can solve the land ownership problem in Assam, we can also minimize the problem of finance to some extent of this sector; which is one of the most important factors for the growth of an industry. The institutional finance and financial assistance from Tea Board of India is totally based on land ownership of the cultivator in Assam.
- In case of marketing, the All Assam Small Tea Growers Association

(AASTGA) can set up co-operative form of cultivation and a standard for production of green leaf and their price before every teagrowing season. For proper price, Government can come to rescue the small tea grower with a minimum support price concept, which is applicable with other agricultural products. Regulation such as adoption of a formula for fixing the price of the tea leaf to the STGs would suffice.

- Looking into the technical problems of the small tea growers, the Tea Board of India, Tea Research Authority and corporate tea producers should take the responsibility to train different aspect of tea cultivation to the small tea growers. This will improve the technical knowhow of the cultivators.
- To restrict the unregulated growth of small tea gardens, Tea Board of India or the Government of India should set up a regulatory authority to look into the uneven growth of this sector. In a liberal economy, a regulator is must for the proper development of a particular sector, who will guide the development process. Recently, the State Government proposes the idea of separate Tea Board for STGs to ensure institutional supports for the growers.

CONCLUSION

The emerging regime of small holding by the growers could be considered a bonus for Indian tea sector. Establishment of small units of production would provide fruitful opportunities for self expression and increasing their material wellbeing of the local rural youths. In Assam, younger generation feel frustrated because they do not have assurance that their education and other skills and talent would be fully utilized. Therefore encouraging these youths in taking up small/ medium level tea cultivation would strengthen the backbone of the rural economy and will also curb social unrest. Economic benefits from the tea industry enjoyed by the British before independence and people from outside the state (Assam) after independence. Native people did not get their due share from the development of tea industry. Tea industry of Assam can be termed as oasis in a desert. The first generation cultivators who have taken up commercial cropping of tea are a radical step towards a new social system in Assam. This will empower the people of Assam with dignity. This silent economic revolution by small tea growers in the rural areas across Assam has helped to tackle unemployment problem apart from using unproductive land in an environment friendly manner. Therefore support should come from every quarter to sustain this socio economic revolution.

Bibliography

Akiyama, Takamasa: *A New Global Tea Model: Specification, Estimation, and Simulation*, World Bank, 1987.

Anderson, Jennifer L.: *An Introduction to Japanese Tea Ritual*, New York: State University of New York Press, 1991.

Anderson, Robert Woodruff: *Tea and Sympathy*, Random House, 1953.

Baildon, Samuel: *The Tea Industry in India*, London: Allen, 1882.

Barker, George: *A Tea Planter's Life in Assam*, Calcutta: Thacker, Spink, 1884.

Blofeld, John, *The Chinese Art of Tea,* London: Allen and Unwin, 1985.

Bramah, Edward. *Tea and Coffee,* London: Hutchinson, 1972.

Crole, David, *Tea,* London: Crosby Lockwood, 1897.

Evans, J.C.: *Tea in China,* New York: Greenwood, 1992.

Forrest, Denys: *Tea for the British,* London: Chatto and Windus, 1973.

Griffiths, Percival Joseph, *The History of the Indian Tea Industry,* London: Weidenfeld & Nicholson, 1967.

Griswold, W.S.: *The Boston Tea Party,* Tunbridge Wells: Abacus, 1973.

Holt, Geraldene, ed.: *A Cup of Tea: An Afternoon Anthology of Fine China and Tea Trades,* New York: Simon and Schuster, 1991.

Iguchi, Kaisen: *Tea Ceremony,* Transl. John Clark. Hoikusha, 1975.

Lillywhite, Bryant: *London Coffee Houses,* London: Allen and Unwin, 1963.

Lu Yü.: *The Classic of Tea: Origins and Rituals,* Transl. and intr. by Francis Ross Carpenter. Hopewell: The Ecco Press, 1974.

Macfarlane, Alan, and Iris Macfarlane: *The Empire of Tea: The remarkable history of the plant that took over the world,* Woodstock: Overlook Press, 2003.

Milton, Ed S.: *Tea: A cultural history from around the world,* Astrolog Publishing House Ltd., 2003.

Moxham, Roy: *Tea: Addiction, Exploitation, and Empire,* New York: Carroll and Graf, 2003.

Okarura, Kakuzo: *The Book of Tea,* Foreword by Elise Grilli. C.E. Turtle Co., 1956.

Pettigrew, Jane: *A Social History of Tea,* London: National Trust, 2001.

Sarkar, Goutam Kumar: *The World of Tea Economy,* Oxford University Press, 1972.

Singh, Shamsher, et. al.: *Coffee, Tea and Cocoa: Market Prospects and Development Lending,* Johns Hopkins University Press, 1977.

Wickizer, Vernon Dale: *Coffee, Tea, and Cocoa: An Economic and Political Analysis,* Stanford University Press, 1951.

Index